D1029386

The complete guide to
British Wildlife

The complete guide to

British Wildlife

Illustrations Norman Arlott

Text Richard Fitter and Alastair Fitter

Collins
London and Glasgow

First published 1981

© in the text Richard Fitter and Alastair Fitter 1981
© in the illustrations Norman Arlott 1981

ISBN Hardback edition 0 00 219224 1
ISBN Paperback edition 0 00 219212 8

Colour reproduction by Adroit Photo-Litho Ltd, Birmingham

Filmset by Jolly & Barber Ltd, Rugby

Printed and bound by Wm Collins Sons and Co. Ltd, Glasgow

Contents

Artist's Preface

When I first approached Collins with a book that would include most of Britain's common wildlife I had little idea of the vast amount of work involved. Drawing up a provisional list of species was not a problem, nor did the prospect of producing some two thousand illustrations seem an insurmountable challenge. However, my knowledge of natural history has its shortcomings and I am an artist rather than writer. It was therefore extremely fortunate that two of the most experienced practitioners and communicators in the field of natural history – Richard and Alastair Fitter – had independently thought of producing such a book, and that this enabled our publishers to bring us together. I would like to record here the gratitude I feel to both of them for the help and constructive criticism they have given me during the preparation of this book. I have learned much.

I must also offer my thanks to Basil Parsons, who accompanied me on numerous expeditions, gave me the benefit of his outstanding knowledge in the field and also applied his trained eye to mistakes in my original drawings. I extend my warmest thanks to Mr and Mrs S. Mactaggart for allowing me to roam free on their Scottish estate on my specimen searches. Thanks are also due on a professional level to John Burton, Michael Chinery and Jim France.

In addition I must thank three people who, though not directly involved in this book, have undoubtedly assisted my progress in the art of natural history illustration. They are Robert Gillmor, Eric Hosking and John G. Williams.

Finally, I must thank my wife and family for their assistance in my search for specimens to illustrate, whether at various times they wanted to or not. To them – Marie, Lisa, Mark and Donna – I dedicate this book.

NORMAN ARLOTT

Abbreviations

♂	male	W	winter
♀	female	L	length
imm.	immature	WS	wing-span
ad.	adult	H	height
S	summer	SH	shoulder height

Introduction

This book aims as far as possible to include within one pair of covers every animal and plant that the average countrygoer or visitor to the seashore is likely to see and want to recognise. Of course there are innumerable invertebrates and lower plants that it was not possible to include, but all the vertebrates (mammals, birds, amphibians, reptiles and fish) that you are likely to see on an ordinary country walk are here, together with all the trees and shrubs and all the more important flowering plants. The insects and other invertebrates, however, and the lower plants (mosses, lichens, fungi and algae) have had to be much more rigorously selected. The ones that are included are those that are most often encountered and are conspicuous enough to arouse the curiosity of people who do not regard themselves as naturalists. Those who are naturalists will possess the appropriate field guide for their own special area of interest, but it is hoped that ornithologists, for instance, will find the sections on trees and wild flowers useful, and that butterfly buffs may learn a thing or two about birds and the plants that provide the food supply of their butterflies' caterpillars. There is an inevitable concentration on the animals and plants of the southern half of Britain, if only because that is where the majority of the population lives. No two people would ever make the same selection – this one is offered in the full knowledge that you cannot put a quart into a pint pot.

The book starts with the vegetable kingdom, since without plants animals could not survive or could not have evolved to their present-day diversity. The most obvious plants by far are the trees, so they begin the book. Then come the rest of the flowering plants, followed by the lower plants that have no flowers but reproduce by means of spores and not seeds. In the evolutionary scale, of course, the lower plants came first, just as invertebrates preceded vertebrates among the animals. However, again that section starts with the most noticeable animals, the mammals. Deer, seals and whales are larger than any bird, though on an ordinary country walk you will be lucky if your list of mammals includes more than the Rabbit and Grey Squirrel, but you should manage to see 20 or 30 different species of birds. After dealing with the three groups of lower vertebrates, we turn to the invertebrates and rather arbitrarily describe first the most popular and conspicuous insects, the butterflies and moths. The animals and the plants of the seashore are dealt with separately after the other invertebrates, since the marine environment is so different from the terrestrial one. Finally there are a few pages on those essential ingredients of the countryside, farm animals and crops.

Since both animals and plants, within the major groupings, have been arranged here in systematic scientific order, a few words about this are necessary. Although the fungi were once considered plants, they are now accorded the status of a separate kingdom of their own, distinct from the plant kingdom. Flowering plants are divided first into monocotyledons and dicotyledons (see p. 99), and then into families. Within each family there are one or more genera, and within each genus one or more species. Thus the common daisy belongs to the Composite Family, and is called *Bellis* (genus) *perennis* (species). There is in fact another species of daisy found in Europe, called *Bellis annua*. In this case, but by no means always, the scientific name gives an important distinction between the two species – one is perennial, the other annual.

The classification of animals differs somewhat from that of plants. Animals are arranged first of all in a number of major groupings called phyla. The vertebrates constitute a sub-group of the phylum Chordata, which also includes three groups of marine creatures, the acorn worms or enteropneusts, the tunicates or sea-squirts, and the lancelets, primitive fish-like creatures found on sandy sea-bottoms. The remaining 21 phyla comprise the invertebrates, of which the principal groupings in this book are the molluscs and the arthropods, together with a number whose members are found only in or at the edge of the sea. Most of the invertebrates that come to the ordinary countrygoer's notice are arthropods, which include the insects (butterflies, moths, bees, wasps, ants, flies, beetles and dragonflies), the crustaceans (lobsters, crabs, shrimps, crayfish and woodlice) and the arachnids (spiders, harvestmen and mites).

Each phylum is subdivided into classes: mammals, birds, reptiles and amphibians are all separate classes, but there are several classes of fish. Within the arthropod phylum the insects, crustaceans and arachnids are all distinct classes. Each class is subdivided into orders, as, for instance, mammals into bats (Chiroptera), carnivores (Carnivora) and odd-toed ungulates (Artiodactyla), and each class into families – among the families of carnivores are the cats (Felidae), dogs (Canidae) and stoats (Mustelidae). From this point the classification is the same as in plants, with further subdivisions into genera and species. Thus the House Sparrow belongs to the class Aves (birds), the order Passeriformes (song-birds), the family Passeridae (sparrows), the genus *Passer* and the species *domesticus*, while in the same way the Fox is in the class Mammalia, the order Carnivora, the family Canidae, the genus *Vulpes* and the species *vulpes*. With animals, unlike plants, the genus and species name may be the same.

A note on sizes

The layout of this book has meant that it was not possible to impose a uniform scale on the illustrations, even with similar groups such as mammals or wild flowers. Instead, for animals the length is given, or sometimes a range of lengths – since many animals are variable – of the average adult. If a particular species has a long beak or snout or tail, this is included in the total length. Where appropriate, a shoulder height is given. In some cases a wing-span is given. Unless otherwise stated in the text, however (see abbreviations on p. 6), measurements are always lengths.

For all the plants, except for the trees, the height of normal, mature specimens is given on the simple scale low–short–tall. These correspond approximately to plants up to about 20 cm high, 20–50 cm, and 50–100 cm respectively. Extremely small plants, less than 5 cm, and extremely large plants, more than 1 m high, are designated very low and very tall respectively.

Trees and Climbers

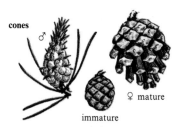

cones

♂

♀ mature

immature

Trees are normally regarded as plants with a single, woody stem more than 3–4 m high which is self-supporting. This book includes all the common native trees and several of the commoner planted species. Being so large, trees can often be identified from their leaves, trunks and general shape, but all trees produce flowers or cones as well. Conifers are the most primitive trees and do not have proper flowers: the pollen and eggs are produced in cones, and the female cones may be large and woody, as in the pines, when they have ripe seed. Other trees are Angiosperms, and these all have true flowers, which may be like everyone's general idea of a flower, as in Horse Chestnut or Crab Apple, or form a catkin, as in Hazel or Oak.

the commonest shrub. On windy or exposed sites, Sea Buckthorn is common, for example, on sand-dunes, and Blackthorn forms scrub on sea-cliffs. The best place to find shrubs is in hedges, a man-made habitat, and it seems that the older the hedge the more shrubs will grow in it.

Climbers rely on something taller than themselves up which to scramble, so they tend to be found in woods and in scrub, particularly in hedges.

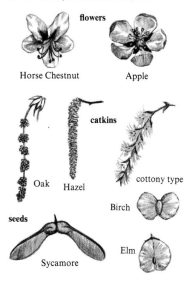

flowers

Horse Chestnut   Apple

catkins

Oak   Hazel

cottony type

Birch

seeds

Sycamore

Elm

fruit

Holly berry   beechnut

Cherry

hazelnut   acorn

leaves

Hazel

Hawthorn

Sea Buckthorn

Blackthorn

Catkins allow pollen to be dispersed by wind. Similarly many trees have wind-dispersed seeds, which may either be cottony, as in Willows and Poplars, or have a special wing which helps them stay airborne, as do Birches, Elms and Sycamore. Others rely on edible fruits for seed dispersal, whether these are fleshy and soft (cherries, Holly) or a hard nut (Hazel, Oak, Beech).

Many woody plants are much smaller than trees. **Shrubs** tend to grow either as the understorey in woods, e.g. Hazel, or as the first colonists of open areas left ungrazed or unmown; Hawthorn is then

9

PINE FAMILY Pinaceae

Douglas Fir *Pseudotsuga menziesii*. A tall evergreen conifer growing to over 50 m in Britain, where it is always planted, and to over 100 m in its native NW America. It has a narrow, open crown with spreading branches, and thick, deeply fissured, purplish brown bark. It has flat needles and highly distinctive cones, which hang from the tips of the branches and have long, 3-toothed tongues projecting from each scale.

Norway Spruce *Picea abies* (1) has a regular conical crown and stiff, sharp needles, which form a distinct peg when they fall. It has long, cylindrical cones which hang from the branches. Norway Spruce is the true Christmas tree and is widely planted for that purpose, but the commonest spruce now is the similar **Sitka Spruce** *P. sitchensis*, from western N America, which has flattened, bluish needles and smaller, rounded cones.

European Larch *Larix decidua* (2). The commonest winterbare conifer, along with Japanese and hybrid larches. It has grey–brown, fissured bark and limp, bright green needles in tufts. The male cones are yellow, the female purplish initially, ripening to a small, smooth, egg-shaped cone. Less frequently planted now than **Japanese Larch** *L. kaempferi*, which has orange–brown twigs and out-turned cone scales, and the intermediate *L. × eurolepis*, the hybrid between them.

♀ cone   ♂ cone

10

Grand Fir *Abies grandis* (1). An impressive and widely planted tree with smooth grey–brown bark on young trunks. It has shiny green leaves projecting almost horizontally from the twigs, and reddish brown, egg-shaped cones which sit upright on the branches. **Common Silver Fir** *A. alba* has stubbier leaves which leave the shoot in all directions except upwards, and larger, cylindrical cones.

Scots Pine *Pinus sylvestris* (2). This, the only native British pine, has cracked bark, brown at the base but orange–brown higher up the trunk. Its needles are in pairs and the yellow male ones cluster round the shoot tips in spring. The female cones are small, green at first, but maturing to a greyish brown. Scots Pine is common in Scotland and widely naturalised elsewhere. **Corsican Pine** *P. nigra* is widely planted; it has longer, dark green needles and a neater crown.

Western Hemlock *Tsuga heterophylla*. An elegant and widely planted tree with a very characteristic drooping leading shoot. It has flat needles in 3 different sizes, giving the shoots an untidy appearance, and small, greenish brown, egg-shaped cones. The male cones are bright red–purple. Western Hemlock is planted for forestry as an undercrop because it can withstand considerable shade.

♂ cone

♀ cone

11

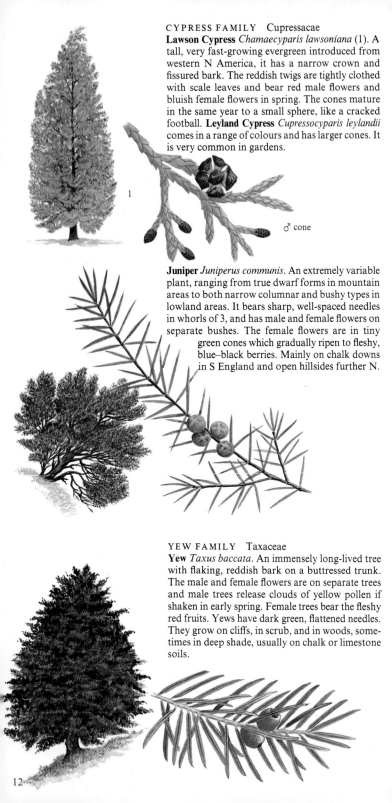

CYPRESS FAMILY Cupressacae

Lawson Cypress *Chamaecyparis lawsoniana* (1). A tall, very fast-growing evergreen introduced from western N America, it has a narrow crown and fissured bark. The reddish twigs are tightly clothed with scale leaves and bear red male flowers and bluish female flowers in spring. The cones mature in the same year to a small sphere, like a cracked football. **Leyland Cypress** *Cupressocyparis leylandii* comes in a range of colours and has larger cones. It is very common in gardens.

♂ cone

Juniper *Juniperus communis*. An extremely variable plant, ranging from true dwarf forms in mountain areas to both narrow columnar and bushy types in lowland areas. It bears sharp, well-spaced needles in whorls of 3, and has male and female flowers on separate bushes. The female flowers are in tiny green cones which gradually ripen to fleshy, blue–black berries. Mainly on chalk downs in S England and open hillsides further N.

YEW FAMILY Taxaceae

Yew *Taxus baccata*. An immensely long-lived tree with flaking, reddish bark on a buttressed trunk. The male and female flowers are on separate trees and male trees release clouds of yellow pollen if shaken in early spring. Female trees bear the fleshy red fruits. Yews have dark green, flattened needles. They grow on cliffs, in scrub, and in woods, sometimes in deep shade, usually on chalk or limestone soils.

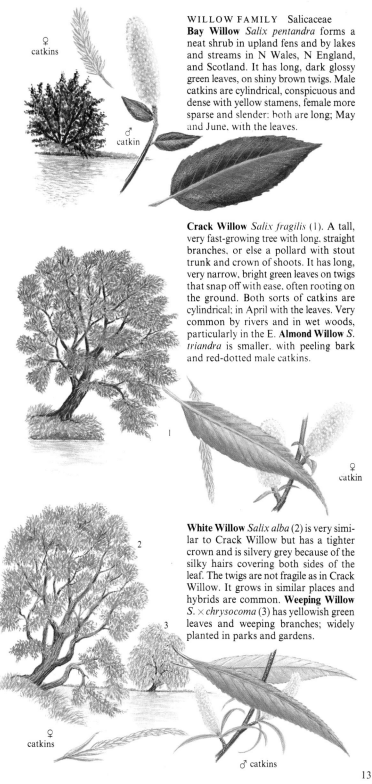

WILLOW FAMILY Salicaceae
Bay Willow *Salix pentandra* forms a neat shrub in upland fens and by lakes and streams in N Wales, N England, and Scotland. It has long, dark glossy green leaves, on shiny brown twigs. Male catkins are cylindrical, conspicuous and dense with yellow stamens, female more sparse and slender; both are long; May and June, with the leaves.

♀
catkins

♂
catkin

Crack Willow *Salix fragilis* (1). A tall, very fast-growing tree with long, straight branches, or else a pollard with stout trunk and crown of shoots. It has long, very narrow, bright green leaves on twigs that snap off with ease, often rooting on the ground. Both sorts of catkins are cylindrical; in April with the leaves. Very common by rivers and in wet woods, particularly in the E. **Almond Willow** *S. triandra* is smaller, with peeling bark and red-dotted male catkins.

1

♀
catkin

White Willow *Salix alba* (2) is very similar to Crack Willow but has a tighter crown and is silvery grey because of the silky hairs covering both sides of the leaf. The twigs are not fragile as in Crack Willow. It grows in similar places and hybrids are common. **Weeping Willow** *S. × chrysocoma* (3) has yellowish green leaves and weeping branches; widely planted in parks and gardens.

2

3

♀
catkins

♂ catkins

Sallow or **Pussy Willow** *Salix caprea* (1) forms a large bush or small tree. Its broad oval leaves are dark green and wrinkled above, greyish underneath; oval yellow male catkins appear in March and April. Female catkins are longer and both appear before the leaves. Very common in woods and scrub. **Grey Willow** *S. cinerea* has narrower leaves and hairy twigs; wetter places. **Eared Willow** *S. aurita* is smaller.

Osier *Salix viminalis* is a much-branched shrub with long, straight branches. It has extremely long (up to 25 cm) narrow leaves, dark green on the top side and silky grey underneath. The male catkins are oval and yellow and appear in April before the leaves, but the longer, cylindrical female catkins appear a little later, often as the leaves expand. Osiers are common by streams and in wet places.

Creeping Willow *Salix repens* (2). A dwarf shrub with a creeping underground stem from which arise woody, leafy stems which may be prostrate or rise to 1–1½ m. It has oval leaves which are silky on both sides when young, as are the twigs. The catkins are small and narrow oval, appearing in April and May after the leaves. It grows in wet heaths and dune-slacks, except in the Midlands. **Least Willow** *S. herbacea* is the commonest of a group of tiny mountain willows only a few cm high. It has round, shiny green leaves.

Grey Poplar *Populus canescens* (1) has a spreading crown and suckers. Its rounded oval leaves have teeth and are greyish below. Long, reddish male catkins and stout, greenish female catkins appear in March. Native in damp woods in the SW; planted elsewhere. **White Poplar** *P. alba* has whiter leaves. Widely planted.

Italian Poplar *Populus* ×*canadensis* (2) is a massive tree with spreading branches and dark, furrowed bark. It bears male catkins only. It is very widely planted and much commoner than the true native **Black Poplar** *P. nigra*, whose branches arch downwards and whose bark bears massive bosses. It occurs throughout England, mainly in the SE. The related **Lombardy Poplar** *P. nigra* var. *italica* (3) has a spire-like crown.

Aspen *Populus tremula* (4) is so named because its rounded leaves flutter in the breeze. It is usually a small tree surrounded by a thicket of suckers, which bear larger, more heart-shaped and toothed leaves. Silky male catkins and purple and green female catkins appear in February and March. In damp woods and on moors throughout Britain. **Balm of Gilead** *P.* ×*gileadensis* has triangular leaves and sticky buds; widely planted.

BIRCH FAMILY Betulaceae

Silver Birch *Betula pendula*. An elegant, slender tree with a white trunk, dark at the base and with diamond-shaped bosses. The branches droop at their tips and have shiny twigs. It has pointed oval leaves with long and short teeth, and conspicuous drooping male catkins in April and May. The female catkins are small and upright but hang when ripe and shed masses of winged seeds. Very common on dry soils.

♀ catkins

♂ catkins

♀ catkins

♂ catkins

Alder *Alnus glutinosa* has a regular, conical crown when young, but mature trees become more open and straggling. It has rounded leaves somewhat like Hazel, except that the tip is usually slightly indented. The catkins are borne on the old wood, the male with purple scales and yellow flowers, the female small and purple –brown in early spring, but maturing to a hard brown cone. Alder is a very common waterside tree and forms woods on lowland peats.

cones

Sweet Gale or **Bog Myrtle** *Myrica gale* is a woody shrub, usually less than head-high, with upright red–brown stems bearing narrow, greyish leaves, widest near their tips. The orange–brown male catkins and the rather squat, reddish female catkins appear before the leaves in April, on different plants, but individual plants may switch sex from year to year. The plant is strongly aromatic, particularly in flower. Common in wet peaty places in the N, but more scattered in the S.

Hazel *Corylus avellana* is a many-stemmed shrub, occasionally forming a small tree in woodland. The zigzag twigs are covered with reddish brown hairs, and the toothed, rounded leaves end in a small point. The male flowers, in hanging catkins, are produced from January to March; the female are small and round with bright red styles. The fruit has a leafy husk and ripens in late autumn. Common in woods and hedges.

fruit

♀
catkin

♀
catkins

Hornbeam *Carpinus betulus*. An attractive tree with a spreading, oval crown, and smooth grey–brown trunk, buttressed at the base. It has toothed, pointed oval leaves with prominent veins, and short, greenish, rather feathery male catkins. The female catkins are short and green in flower, in April and May, but ripen to a dangling cluster of 3-lobed, winged fruits. Mainly found in SE England where it forms woods with Oak, but often planted elsewhere.

♀
catkin

WALNUT FAMILY   Juglandaceae
Walnut *Juglans regia*. A magnificent tall tree with a spreading crown and smooth, grey bark. The greenish twigs bear large, black buds and dull, yellowish green leaves with 7–9 pairs of pointed leaflets. Each tree bears long, greenish male catkins and inconspicuous female flowers; May to June. Widely planted in hedges and parks.

♀
flowers

♂
flowers

fruit

17

BEECH FAMILY Fagaceae

Beech *Fagus sylvatica*. A spreading tree with smooth, grey bark and pointed buds. The oval leaves are veined and silky; the male flowers are in clusters on long stalks, with the leaves in April and May. Female flowers are smaller and upright, ripening to nuts. Beech forms pure woods on dry soils in S England.

fruit

♂ catkins

♀ flowers

♂ flowers

Sweet Chestnut *Castanea sativa*. A tall, spreading tree with deeply fissured brown bark, usually twisting spirally round the trunk. It has long, narrow, strongly-toothed leaves and dangling, rather sparse catkins, with yellowish white male flowers at the tip and female at the base. It flowers later than most trees, in July, and the delicious edible fruits, encased in a spiny shell, ripen in October. Introduced and widely planted.

fruit

leaf of 2

1

fruit of 2

Pedunculate Oak *Quercus robur* (1). Very familiar with its broad crown, bent branches, and ridged bark. It has deeply-lobed leaves on tiny stalks, and yellow–green catkins, the male longer than the female. The fruit is the acorn in a scaly, long-stalked cup. Found throughout Britain, it is dominant on heavy lowland soils. **Durmast Oak** *Q. petraea* has stalked leaves and unstalked acorns. Commonest in the W. **Evergreen Oak** *Q. ilex* (2) has almost untoothed evergreen leaves. Planted.

leaf of 1

fruit of 1

18

Horse Chestnut *Aesculus hippocastanum* has a massively domed crown and spreading branches. It has grey–brown bark and large, sticky, brown buds, opening to large palmate leaves with 5–7 leaflets. The flowers are 5-petalled and creamy, with a yellow or pink–orange spot, in tall candelabra in April and May. The fruit is a brown nut in a spiny green case. Introduced and widely planted.

fruit

flower

ELM FAMILY Ulmaceae

English Elm *Ulmus procera* has a tall, narrow crown with arching branches. It has deep dark green, broad oval leaves. The red, tufted flowers are borne on bare branches in February and March, ripening by June to a winged fruit with the seed above the middle. Once widespread in S England, now ravaged by disease. **Small-leaved Elm** *U. minor* has narrower, lighter green leaves.

1

fruit

flowers

fruit

2

Wych Elm *Ulmus glabra* (2) has a much more spreading crown than other elms, and bark that is rough in old trees. It produces no suckers. The rough-surfaced leaves have very short stalks and the fruit has the seed below the middle. Wych Elm is a common woodland tree, particularly in the N and W, and is more resistant to Dutch Elm Disease than other elms.

flowers

19

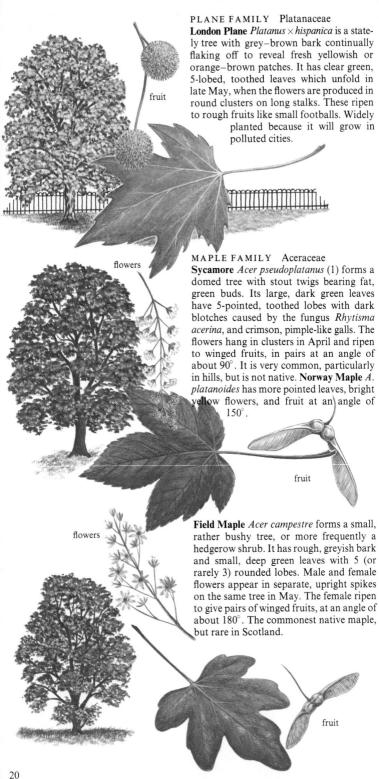

PLANE FAMILY Platanaceae
London Plane *Platanus × hispanica* is a stately tree with grey–brown bark continually flaking off to reveal fresh yellowish or orange–brown patches. It has clear green, 5-lobed, toothed leaves which unfold in late May, when the flowers are produced in round clusters on long stalks. These ripen to rough fruits like small footballs. Widely planted because it will grow in polluted cities.

fruit

MAPLE FAMILY Aceraceae
Sycamore *Acer pseudoplatanus* (1) forms a domed tree with stout twigs bearing fat, green buds. Its large, dark green leaves have 5-pointed, toothed lobes with dark blotches caused by the fungus *Rhytisma acerina*, and crimson, pimple-like galls. The flowers hang in clusters in April and ripen to winged fruits, in pairs at an angle of about 90°. It is very common, particularly in hills, but is not native. **Norway Maple** *A. platanoides* has more pointed leaves, bright yellow flowers, and fruit at an angle of 150°.

flowers

fruit

Field Maple *Acer campestre* forms a small, rather bushy tree, or more frequently a hedgerow shrub. It has rough, greyish bark and small, deep green leaves with 5 (or rarely 3) rounded lobes. Male and female flowers appear in separate, upright spikes on the same tree in May. The female ripen to give pairs of winged fruits, at an angle of about 180°. The commonest native maple, but rare in Scotland.

flowers

fruit

HOLLY FAMILY Aquifoliaceae
Holly *Ilex aquifolium* is a small tree or shrub, often with branches down to the ground. It has green twigs and very dark, glossy green leaves with spiny, wavy edges; some of the leaves, particularly high up, may be smooth-edged. The small, white, 4-petalled flowers appear in short, stiff clusters from May to August, and ripen to red berries. Common in hedges and woods and on rocks.

flowers

fruit

lower leaf

top leaf

LIME FAMILY Tiliaceae
Common Lime *Tilia × vulgaris* (1) is very tall with a ridged, bossed trunk and often many suckers. It has large, heart-shaped leaves and flowers in clusters of 5–10; July. Widely planted, often in avenues, it originated as a hybrid between **Large-leaved Lime** *T. platyphyllos*, with larger leaves, unbossed trunk and flowers in threes, and **Small-leaved Lime** *T. cordata*, with small, birch-like leaves, bluish underneath. Both are uncommon native woodland trees.

1

fruit

flowers

OLIVE FAMILY Oleaceae
Ash *Fraxinus excelsior* has a ragged crown, often dying back at the top, although neat and narrow when young. It has grey twigs with large, black buds, and pinnate leaves with 3–7 pairs of leaflets. It comes into leaf after the flowers, which open in April. Each fruit is set in an oblong wing, hanging in bunches. A woodland tree, particularly on limestone, and very common in N England.

fruit

flowers

ROSE FAMILY   Rosaceae

Rowan *Sorbus aucuparia* is slender with steeply rising branches. It has smooth, greyish bark with hairy buds, and pinnate leaves. The creamy, 5-petalled flowers, 5–10 mm, appear in flat-topped clusters in May and ripen to scarlet berries by August. Scattered in woods, and on heaths and moors.

Whitebeam *Sorbus aria* (1) forms a spreading tree with a characteristic silvery appearance, especially in wind when the silvery grey undersides of the toothed, pointed oval leaves are exposed. The flowers, 10–15 mm across, appear in dense, flat-topped clusters in May and June; the fruits are scarlet and slightly oval. Scattered in woods and scrub on limestone, but mainly in S England. **Swedish Whitebeam** *S. intermedia*, with shallowly-lobed leaves, is widely planted and sometimes bird-sown.

Wild Pear *Pyrus communis*. A small tree with an untidy crown. The spiny twigs bear numerous 'short shoots' which produce the leaves, flowers and fruit. It has hairless, rounded leaves and carries clusters of 5-petalled, white flowers in April, as the leaves emerge and before the similar Crab Apple (p. 23). Although this is the parent of all cultivated pears, the fruit is small (to 4 cm), brown, hard, and often rounded. A widespread but not common hedgerow and woodland tree, from N England southwards.

22

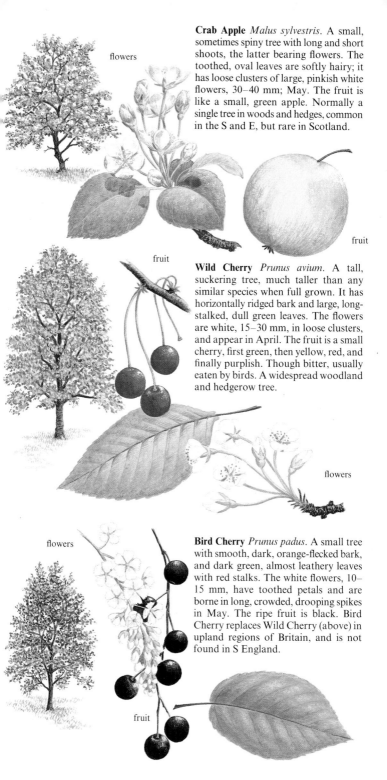

Crab Apple *Malus sylvestris*. A small, sometimes spiny tree with long and short shoots, the latter bearing flowers. The toothed, oval leaves are softly hairy; it has loose clusters of large, pinkish white flowers, 30–40 mm; May. The fruit is like a small, green apple. Normally a single tree in woods and hedges, common in the S and E, but rare in Scotland.

flowers

fruit

fruit

Wild Cherry *Prunus avium*. A tall, suckering tree, much taller than any similar species when full grown. It has horizontally ridged bark and large, long-stalked, dull green leaves. The flowers are white, 15–30 mm, in loose clusters, and appear in April. The fruit is a small cherry, first green, then yellow, red, and finally purplish. Though bitter, usually eaten by birds. A widespread woodland and hedgerow tree.

flowers

flowers

Bird Cherry *Prunus padus*. A small tree with smooth, dark, orange-flecked bark, and dark green, almost leathery leaves with red stalks. The white flowers, 10–15 mm, have toothed petals and are borne in long, crowded, drooping spikes in May. The ripe fruit is black. Bird Cherry replaces Wild Cherry (above) in upland regions of Britain, and is not found in S England.

fruit

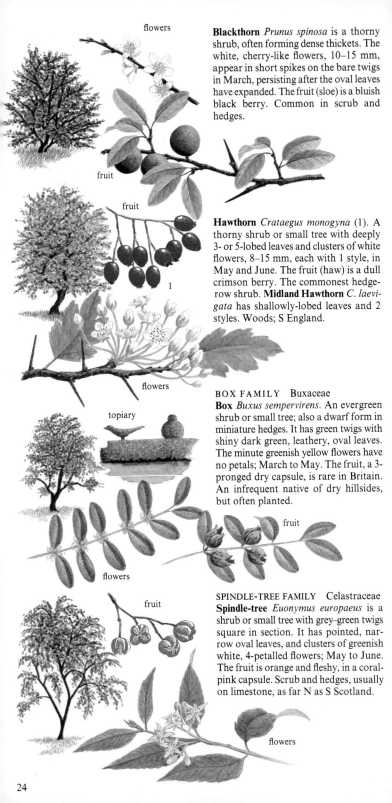

Blackthorn *Prunus spinosa* is a thorny shrub, often forming dense thickets. The white, cherry-like flowers, 10–15 mm, appear in short spikes on the bare twigs in March, persisting after the oval leaves have expanded. The fruit (sloe) is a bluish black berry. Common in scrub and hedges.

flowers

fruit

fruit

1

flowers

Hawthorn *Crataegus monogyna* (1). A thorny shrub or small tree with deeply 3- or 5-lobed leaves and clusters of white flowers, 8–15 mm, each with 1 style, in May and June. The fruit (haw) is a dull crimson berry. The commonest hedgerow shrub. **Midland Hawthorn** *C. laevigata* has shallowly-lobed leaves and 2 styles. Woods; S England.

BOX FAMILY Buxaceae

Box *Buxus sempervirens*. An evergreen shrub or small tree; also a dwarf form in miniature hedges. It has green twigs with shiny dark green, leathery, oval leaves. The minute greenish yellow flowers have no petals; March to May. The fruit, a 3-pronged dry capsule, is rare in Britain. An infrequent native of dry hillsides, but often planted.

topiary

fruit

flowers

SPINDLE-TREE FAMILY Celastraceae

Spindle-tree *Euonymus europaeus* is a shrub or small tree with grey–green twigs square in section. It has pointed, narrow oval leaves, and clusters of greenish white, 4-petalled flowers; May to June. The fruit is orange and fleshy, in a coral-pink capsule. Scrub and hedges, usually on limestone, as far N as S Scotland.

fruit

flowers

24

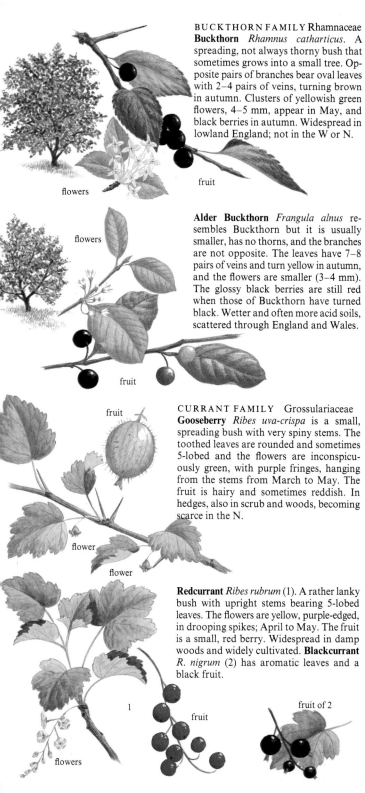

BUCKTHORN FAMILY Rhamnaceae
Buckthorn *Rhamnus catharticus*. A spreading, not always thorny bush that sometimes grows into a small tree. Opposite pairs of branches bear oval leaves with 2–4 pairs of veins, turning brown in autumn. Clusters of yellowish green flowers, 4–5 mm, appear in May, and black berries in autumn. Widespread in lowland England; not in the W or N.

flowers

fruit

Alder Buckthorn *Frangula alnus* resembles Buckthorn but it is usually smaller, has no thorns, and the branches are not opposite. The leaves have 7–8 pairs of veins and turn yellow in autumn, and the flowers are smaller (3–4 mm). The glossy black berries are still red when those of Buckthorn have turned black. Wetter and often more acid soils, scattered through England and Wales.

flowers

fruit

CURRANT FAMILY Grossulariaceae
Gooseberry *Ribes uva-crispa* is a small, spreading bush with very spiny stems. The toothed leaves are rounded and sometimes 5-lobed and the flowers are inconspicuously green, with purple fringes, hanging from the stems from March to May. The fruit is hairy and sometimes reddish. In hedges, also in scrub and woods, becoming scarce in the N.

fruit

flower

flower

Redcurrant *Ribes rubrum* (1). A rather lanky bush with upright stems bearing 5-lobed leaves. The flowers are yellow, purple-edged, in drooping spikes; April to May. The fruit is a small, red berry. Widespread in damp woods and widely cultivated. **Blackcurrant** *R. nigrum* (2) has aromatic leaves and a black fruit.

1

flowers

fruit

fruit of 2

OLEASTER FAMILY Elaeagnaceae
Sea Buckthorn *Hippophae rhamnoides*. A large shrub or small tree which produces suckers and forms dense thickets. It has brown, thorny twigs and long, narrow, silvery leaves. The tiny green flowers are produced, male and female, on separate plants in April and May; only female plants bear the bright orange berries. On dunes all around the coast and often planted inland.

flowers

fruit

ROSE FAMILY Rosaceae
Cherry Laurel *Prunus laurocerasus* is indeed a cherry, and its flowers and fruit are similar to Bird Cherry (p. 23). It is a large, spreading shrub or small tree with long, glossy, stiff and rather brittle leaves. The flowers, 8–10 mm across, appear in upright spikes in April and May, and the fruit is an oblong, shiny purplish black cherry. It is introduced but often naturalised in woods.

flowers

fruit

HEATH FAMILY Ericaceae
Rhododendron *Rhododendron ponticum*. Often confused with Cherry Laurel (above), but the leaves, flowers and fruit are quite different. The flowers appear in dense clusters at the tips of the branches in June, each shaped like a flattened bell up to 60 mm across. The seeds are in a dry capsule. Widely naturalised in woods.

flower

BUDDLEIA FAMILY Buddlejaceae
Buddleia *Buddleia davidii*. A common garden shrub, also known as Butterfly Bush because of the numbers of butterflies attracted to its purple flowers from June to October. It has long, narrow, dark or greyish green leaves on stout, angled, pithy twigs. Naturalised in waste places, mainly in S England, and increasing.

flower-spike

DOGWOOD FAMILY Cornaceae
Dogwood *Cornus sanguinea*. A strikingly red-stemmed shrub, usually producing a mass of slender stems, reaching up to 4 m. The oval leaves have 3–5 pairs of prominent leaves and turn scarlet in autumn. Dogwood bears dense clusters of small 4-petalled flowers, 8–10 mm, from May to July, ripening to black berries. Common in hedges and scrub in the S, often on chalk; rare in the N and only introduced in Scotland.

OLIVE FAMILY Oleaceae
Privet *Ligustrum vulgare* (1) is a tall shrub, reaching 5 m; it loses its leaves very late. The downy twigs have opposite pairs of narrow, deep glossy green leaves, and end in branching spikes of 4-petalled flowers, 4–6 mm; May to June. The fruit is a glossy black berry. Common in scrub, usually on chalk or limestone, becoming rare in Scotland. **Garden Privet** *L. ovalifolium*, the urban hedging plant, has broader leaves.

HONEYSUCKLE FAMILY Caprifoliaceae
Snowberry *Symphoricarpos rivularis* is an introduced shrub that forms dense thickets 1–2 m high. The main stems bear rounded leaves but it suckers freely and those shoots may have lobed leaves. The tiny pink flowers appear from June to September and ripen to a dull white berry in late autumn. Snowberry is widely planted and often naturalised.

Elder *Sambucus nigra* forms a shrub or a small tree, occasionally up to 10 m high, with deeply furrowed bark. The smooth, pithy twigs carry pinnate leaves with 2–3 pairs of leaflets and dense, upright, flat-topped clusters of flowers, 10–20 cm across, in June. In fruit the weight of the black berries bends the red stalks down. Extremely common, except in mountains, it springs up like a weed on bare ground and paths.

Guelder Rose *Viburnum opulus*. A large shrub, 2–4 m high, with 3–5-lobed leaves, whose lobes are themselves deeply toothed. The flowers are in spreading, flat-topped clusters, the inner ones only 5–6 mm across, the outer ones 15–20 mm, with very unequal petals, the largest on the outside; May to July. The fruit is a round red berry. Common in scrub, usually on damp soils, even in fens, becoming scarce in Scotland.

Wayfaring Tree *Viburnum lantana* is a large shrub, 2–4 m and even higher, which has powdery twigs and evenly-toothed oval leaves. The fragrant flowers are in a head similar to Guelder Rose (above) but all equal in size; April to June. The fruit is a flattened oval black berry. Mainly in chalk scrub in S England; introduced elsewhere.

27

Honeysuckle *Lonicera periclymenum* is a clockwise-twining climber with flaking, woody stems up to 6 m high. It has oval leaves on very short stalks and almost circular heads of tubular, 2-lipped flowers, wonderfully fragrant from June to September. The tube is up to 50 mm long. The red berries are in tight clusters. Common in woods and hedges.

HEMP FAMILY Cannabaceae
Hop *Humulus lupulus* is a clockwise-twining climber with woody stems that can scramble up to 6 m up trees and shrubs. It has large, 5-lobed leaves and open, branching clusters of greenish male flowers in July and August. The female flowers are in a green 'cone' and when ripe form the hops that flavour beer. Common in hedges, scrub, and damp woods; scarce in the W and rare in Scotland.

BUTTERCUP FAMILY Ranunculaceae
Old Man's Beard *Clematis vitalba* is a scrambling, woody perennial found all over trees and in thickets. Its pinnate leaves have 3–5 leaflets with twining stalks, and the fragrant, creamy flowers, 20 mm, open in July and August and ripen to clusters of fruits with long, hairy plumes. Common in scrub on chalk and limestone in the S, rare N of the Humber–Mersey.

GOURD FAMILY Cucurbitaceae
White Bryony *Bryonia cretica* is a tall perennial which climbs by long, unbranched tendrils arising from the base of the 5-lobed leaves, as do the clusters of greenish flowers, 10–15 mm across. Male and female flowers are on separate plants; May to September. The fruit is a shiny red berry; in S, E, and central England.

IVY FAMILY Araliaceae
Ivy *Hedera helix* is woody and climbs walls and tall trees by tiny roots, or carpets the ground in woods. It has glossy 3- or 5-lobed leaves, often conspicuously veined. Flowering shoots are only produced in sun and have unlobed leaves and clusters of green flowers, September to November, ripening to rounded, flat-topped, black berries. Common except in mountains.

YAM FAMILY Dioscoraceae
Black Bryony *Tamus communis* is a tall, clockwise-climbing perennial with herbaceous stems and no tendrils. It has well-veined, heart-shaped leaves and long spikes of 6-petalled, yellow–green flowers, 4–5 mm, male and female on separate plants; May to August. The red berry is often yellowish or greenish. The only native yam, it is common in the S, N to the Lake District.

Wild Flowers

'Wild flowers' is a vague term, without scientific meaning, and yet conveying a useful image to most people. Flowering plants is the precise grouping of all plants that have true flowers, excluding ferns and their allies which reproduce by spores, and even conifers (strictly the Gymnosperms), which have cones, but not the fully developed structure we know as a flower.

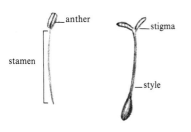

stamen — anther — stigma — style

A flower may contain either male or female reproductive organs, or more usually both. The male organs are the anthers, which are borne on filaments, the 2 together making a stamen. Inside the anther are produced the pollen grains, which are really a sort of spore. They fall on or are carried to the stigma, the receptive surface of the style, the visible part of the female organ. The pollen grain germinates, like a spore of a fern would, and grows down inside

the style to fertilise the egg. The really distinctive feature of a flower is the fact that the egg is totally concealed inside an ovary.

Many trees, but not conifers, and all the grasses, sedges, rushes, and water-weeds, which are dealt with in separate sections of this book, are therefore true flowering plants, for they have flowers. A flower need not be a showy, colourful structure with conspicuous petals and sepals. These are devices which serve to attract insects to the flower, which

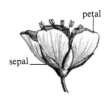

petal — sepal

unwittingly carry the pollen to other flowers, though their motive is to collect nectar, not to pollinate flowers. Some plants are pollinated by wind and for them colourful petals would serve no purpose; they have usually inconspicuous flowers, often gathered into catkins which hang in the wind, and often with very long anther filaments, to distribute the pollen, and styles, to catch it.

Often, however, you will want to know something of a plant that has no flowers, for they are usually present for only a few weeks or months, and here the leaves are important features. Leaves vary from long and narrow, as in the grasses, through spear shapes, ellipses, ovals, and near circles. In addition their edges may be smooth, toothed, or they may be deeply cut in individual leaflets, 3 (trefoil), 5 (palmate), or many (pinnate). Generally plants in shade have large, undivided leaves, whereas those in dry, sunny places have narrow, often dissected leaves, which are better at staying cool.

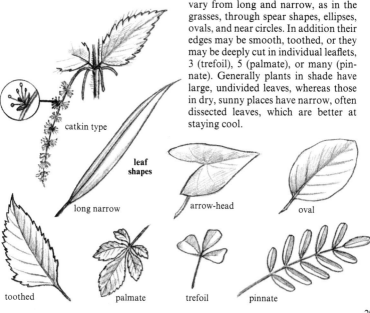

catkin type

leaf shapes

long narrow

arrow-head

oval

toothed

palmate

trefoil

pinnate

MISTLETOE FAMILY Loranthaceae
Mistletoe *Viscum album* is an evergreen woody parasite on the branches of trees. The branches fork regularly and equally and bear pairs of yellowish, leathery, elliptical leaves. The flowers are inconspicuous from February to April, but the fruit is a sticky white berry; November to January. Occurs mainly on Poplar and Apple, almost confined to S and E England.

HEMP FAMILY Cannabaceae
Hemp *Cannabis sativa* is a tall annual whose leaves have 3–9 narrow, toothed lobes, cut almost to the base. Male and female flowers are on separate plants, the male in branching clusters, the female in stalked spikes; July to September. An occasional casual on waste ground, and surreptitiously cultivated for marijuana.

NETTLE FAMILY Urticaceae
Nettle *Urtica dioica* (1) is a familiar tall perennial, copiously endowed with stinging hairs. The short-stalked, toothed leaves are in opposite pairs, and the flowers in long catkins, male and female on separate plants; June to September. In woods and disturbed ground, indicating past human occupation.
Annual Nettle *U. urens* is a smaller annual with long-stalked leaves. An arable weed.

Pellitory of the Wall *Parietaria judaica* is a short or tall, downy perennial with branching red stems, and long-stalked, narrow oval leaves. The flowers are in tight clusters at the base of the leaves; June to October. A characteristic plant of walls and rocks, common in the S, scarce in Scotland.

DOCK FAMILY Polygonaceae have small flowers and sheaths at the leaf-bases. Bistorts *Polygonum* have 5 petals and docks *Rumex* have 6; in both they persist in fruit.
Japanese Knotweed *Reynoutria japonica* is a very tall, fast-growing perennial with thick, zigzag stems. The flowers are borne in branching spikes arising from the base of the huge, triangular leaves; August to October. Introduced from Japan and now an invasive and persistent weed of waste ground, often forming dense thickets.

Knotgrass *Polygonum aviculare* is a low, stiff-stemmed annual with very small leaves on the flowering stems and larger ones (up to 50 mm) on the main branches. The leaf-junctions are surrounded by a silvery sheath. Flowers are white or pale pink in clusters of 1–6; June to November. Very common on bare ground, paths, and seashores.

Bistort *Polygonum bistorta* is a short or tall perennial with upright, unbranched stems bearing oval or almost triangular leaves with winged scales. The flowers are in a dense, cylindrical spike up to 15 mm in diameter; June to August. Typically found in extensive patches in hay meadows; commonest in the Pennines, absent from the Midlands.

Amphibious Bistort *Polygonum amphibium* is a short, creeping perennial when growing on land, with oblong leaves rounded at the base. In water it has broader, untapered, floating leaves. The flowers in both cases are in dense, almost globular, spikes; June to September. Common in lowland regions in and beside still or slow fresh water, and occasionally as a weed.

Alpine Bistort *Polygonum viviparum* is a short, slender, unbranched perennial with narrow leaves. The flower-spike has small, pale pink flowers in the top half and red bulbils in the lower half; June to August. Mountain grassland and rock-ledges, descending to sea-level in N Scotland.

Water Pepper *Polygonum hydropiper* is a short, untidy annual with narrow, unstalked leaves and a burning taste. The flowers are white, tinged with green or pink, in a slender, nodding, leafy spike; July to September. A common plant of wet, usually disturbed, ground.

Redshank *Polygonum persicaria* (1) is a short, occasionally tall, sprawling annual with reddish, well-branched stems. The leaves are narrow, usually have a dark mark, and their short stems are hidden in a fringed sheath. Flowers are in dense spikes; June to October. Very common on disturbed ground. **Pale Persicaria** *P. lapathifolium* is larger and slightly hairy and its flowers are normally greenish white.

Black Bindweed *Bilderdykia convolvulus* is a tall, clockwise-twining annual with pointed heart-shaped leaves. The greenish flowers are like those of the docks *Rumex* and are borne in lax, leafy spikes; July to October. A common and persistent weed of disturbed ground in lowland Britain. The true bindweeds (see p.72) have similar leaves but large, white flowers.

31

Curled Dock *Rumex crispus* is a tall perennial with long, rather narrow, wavy-edged leaves. The flowers are in whorls, densely packed in leafless spikes, none of which spreads away from the main stem; June to October. The fruits have oval segments, each with a prominent brown swelling. A very common plant of disturbed ground, sand-dunes and shingle banks.

Clustered Dock *Rumex conglomeratus* (1) is a tall perennial with spreading branches and a somewhat zigzag stem. The leaves are narrow oblong and the flowers are in whorls in a leafy spike; June to September. All the fruit segments have prominent swellings. Woods, grassy and waste places, becoming scarce in the far N. **Wood Dock** *R. sanguineus* has upright branches, straight stems, a scarcely leafy flower-spike, and only a single swelling on the fruit. Found in woods.

Common Sorrel *Rumex acetosa* (2) is a tall, acid-tasting perennial with arrow-shaped leaves. The reddish orange flowers are in loose spikes; May to July. The segments of the fruit are rounded and have no swellings. A common grassland plant at all altitudes. **Mountain Sorrel** *Oxyria digyna* has fleshy, kidney-shaped leaves; it grows in the mountains.

Sheep's Sorrel *Rumex acetosella* is a slender, low or short perennial with arrow-shaped leaves, the lobes of which are bent forwards. The flower-spikes are almost leafless and have separate male and female flowers; May to August. A very common plant of bare, dry, often sandy or rocky ground, usually on acid soils.

Broad-leaved Dock *Rumex obtusifolius* is a tall perennial with broad oblong leaves up to 25 cm long. The lower leaves are heart-shaped at the base. The flowers are in whorls in a long, loose, branching spike; June to October. The segments of the fruit are toothed and usually only one has a prominent red swelling. A very common plant of disturbed ground.

Water Dock *Rumex hydrolapathum* is a very tall, much-branched perennial with enormous, narrow oval leaves, sometimes over 1 m long. The flowers are in dense, branching spikes; July to September. The segments of the fruit are triangular, each with a prominent swelling. Widespread on the edges of rivers, ditches and ponds in the S and E, becoming scarce further N.

32

Spring Beauty *Montia perfoliata* is a rather fleshy, short annual with long-stalked, oval leaves arising from the stem-base. The flowering stems bear just a pair of unstalked leaves, joined at their bases, beneath the lax spike of white flowers, about 5 mm; April to July. Introduced from western N America and now a widespread plant on dry, often disturbed, soils.

Pink Purslane *Montia sibirica* is a short annual or perennial and has long-stalked, fleshy, pointed oval basal leaves. The 2 unstalked stem-leaves are opposite one another but not joined at their bases. The flowers, 15–20 mm, have notched pink petals with dark veins; April to July. Introduced from western N America and now widely distributed in damp woods, especially near streams.

GOOSEFOOT FAMILY Chenopodiaceae are mainly annuals, with minute, green, petal-less flowers, the male and female ones separate in the Oraches.

Fat Hen *Chenopodium album* is a very variable tall annual with reddish stems and dull green leaves, covered in white meal. The leaves vary from diamond-shaped to very narrow. The flowers are tiny and packed in several leafy spikes; June to October. This is much the commonest *Chenopodium,* and is abundant everywhere on disturbed ground.

Red Goosefoot *Chenopodium rubrum* (1) is a rather variable tall annual with red stems and bright green, coarsely toothed leaves. The flowers are in dense, usually leafy, sometimes branched, spikes; July to October. A common plant of disturbed ground, particularly manure heaps, but rare in Scotland. **Good King Henry** *C. bonus-henricus* has triangular leaves and leafless flower-spikes. It is perennial.

Common Orache *Atriplex patula* is a very variable, tall, branching annual. The lower leaves vary from broadly triangular to narrow, but the upper ones are narrow, tapering on to the stalk. The flower-spikes have separate male and female flowers, but both are tiny and densely crowded; July to October. The fruits are surrounded by spiky bracts. Very common in cultivated ground.

Babington's Orache *Atriplex glabriuscula* is very similar to Common Orache (above) but the leaves are always triangular. The whole plant typically turns red in autumn. It flowers from July to September. Common on bare ground near the sea.

1

33

Sea Beet *Beta vulgaris* ssp *maritima* is a sprawling perennial with tall, reddish shoots. The triangular or diamond-shaped leaves, which may be large, are dark green and leathery. The flowers are in small groups in a long spike, each group at the base of a tiny leaf; July to September. A widespread plant of, usually, sandy seashores, becoming rare in the N. Subspecies *vulgaris* includes cultivated beet, beetroot, etc.

Sea Purslane *Halmione portulacoides* is a short to tall shrub, densely covered with silvery meal. The lower leaves are in opposite pairs, becoming narrower up the stem. The flowers are in dense clusters scattered along branched spikes, and separate male and female flowers occur; July to October. It grows in salt-marshes, typically along the banks of channels, scarcely reaching into Scotland.

Glasswort *Salicornia europaea* is a curious, short, extremely fleshy annual, well-branched, and changing through the season from dark to yellow–green and finally to yellow– or red–brown. The tiny flowers are produced at the stem-junction; August to September. A characteristic plant of salt-marshes, most often near the low-tide mark.

Annual Seablite *Suaeda maritima* is a short, fleshy, much-branched annual varying from greyish to deep red and clothed in short, semi-cylindrical, pointed leaves. The inconspicuous flowers are produced in small groups (1–3) at the base of the leaves from August to October, but ripen to shiny, black seeds. A widespread and common salt-marsh plant, occasionally found on sandy shores.

Prickly Saltwort *Salsola kali* (2) is a sprawling annual that is sometimes hairy and sometimes not. The fleshy, narrow, almost cylindrical leaves have noticeably spiny tips, and the inconspicuous flowers occur singly at the base of the leaves; July to October. Very characteristic of dry, sandy shores.

AMARANTH FAMILY Amaranthaceae **Love Lies Bleeding** *Amaranthus caudatus* (1) is a tall, greyish annual with oval leaves and long, drooping, brilliant red or purple flower-spikes; July to October. It is commonly grown in gardens and is not infrequently found as an escape. Several other amaranths may also escape; most have green flowers.

34

PINK FAMILY Caryophyllaceae have simple leaves in opposite pairs and 4–5-petalled flowers.

Thyme-leaved Sandwort *Arenaria serpyllifolia* is a delicate, thin-stemmed annual with greyish, much-branched stems. The unstalked oval leaves are in well-spaced opposite pairs. The flowers are 5–8 mm across, the sepals longer than the petals; April to November. A common plant of dry, bare places and walls throughout Britain, and often abundant near rabbit burrows.

Spring Sandwort *Minuartia verna* is a low perennial forming tight cushions or loose mats. It has very narrow, slightly curved, 3-veined leaves in whorls and white flowers, 8–10 mm, on shoots rising from the mat; May to September. The petals are longer than the sepals. Locally on rocks, particularly old mine spoil-heaps, in upland areas.

Sea Sandwort *Honkenya peploides* is a very fleshy, creeping perennial with yellow–green, pointed oval leaves in opposite pairs. The flowers are greenish, 6–10 mm; May to August. Most plants have either male or female flowers, the petals as long as the sepals in male flowers but shorter in the female ones. A widespread and common plant of coastal sand and shingle.

Greater Stitchwort *Stellaria holostea* (1) is a short, untidy, square-stemmed perennial with opposite pairs of unstalked, narrow leaves. The flowers are 20–30 mm across, with deeply notched petals and 3 styles; April to June. A very common plant of woods and hedges, usually on heavy soils. **Lesser Stitchwort** *S. graminea* is smaller, with smaller flowers and shorter petals; May to August.

Common Chickweed *Stellaria media* is a very variable, low or short annual with a line of hairs down the stem. It has small, long-stalked lower leaves, and larger, unstalked, sometimes slightly fleshy, oval upper leaves. The flowers are small, less than 10 mm, with thin, bifid petals no longer than the sepals and 3 styles; all year. A very common weed of disturbed ground.

Water Chickweed *Myosoton aquaticum* is a tall, straggling perennial with broad oval or heart-shaped leaves with wavy edges. The flowers are freely produced, 1–15 mm, with stickily hairy stalks; June to October. The petals are deeply split and longer than the sepals, and it has 5 styles. In damp woods and by streams from N England southwards.

1

35

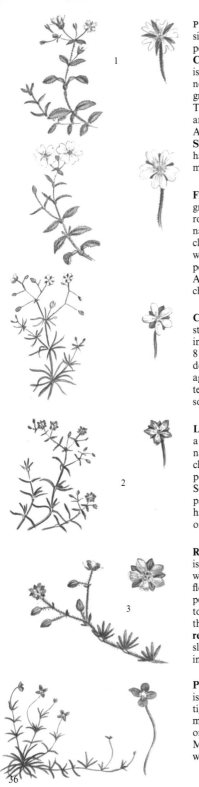

PINK FAMILY Caryophyllaceae have simple leaves in opposite pairs and 4–5-petalled flowers.

Common Mouse-ear *Cerastium fontanum* (1) is a low to short, hairy perennial with leafy non-flowering shoots. The narrow, dark greyish green leaves are in well-spaced pairs. The flowers, 10–15 mm across, have 5 styles, and notched petals longer than the sepals; April to November. Common in grassland. **Sticky Mouse-ear** *C. glomeratum* is a stickily hairy annual with tight heads of flowers. In more open places.

Field Mouse-ear *Cerastium arvense* is a low, greyish, slightly hairy perennial, with long, rooting non-flowering shoots. The leaves are narrow and in pairs at the top of the stem, clusters near the base. It has conspicuous white flowers, 12–20 mm, with notched petals longer than the sepals; April to August. In dry grassland and roadsides on chalky and sandy soils, mainly in the E.

Corn Spurrey *Spergula arvensis* is a short, stickily hairy annual with very narrow leaves in well-spaced whorls. The flowers are 4–8 mm, on long stalks which bend sharply down after flowering, and then turn up again when the fruit is ripe; June to September. A widespread weed of acid, sandy soils.

Lesser Sea Spurrey *Spergularia marina* (2) is a low, fleshy annual with opposite pairs of narrow, slightly pointed leaves. It has loose clusters of rose-pink flowers, 6–8 mm, whose petals are shorter than the sepals; June to September. It is widespread in the drier parts of salt-marshes. **Sand Spurrey** *S. rubra* has greyish leaves and is not fleshy; it grows on inland sandy soils.

Rock Sea Spurrey *Spergularia rupicola* (3) is a well-branched, stickily hairy perennial which appears to have whorls of leaves. The flowers are deep pink, 8–10 mm, with the petals slightly longer than the sepals; June to September. Found on cliffs and rocks by the sea in the S and W. **Greater Sea Spurrey** *S. media* is almost hairless and has slightly purplish flowers. In salt-marshes, in muddier places than Lesser Sea Spurrey.

Procumbent Pearlwort *Sagina procumbens* is a low, hairless, creeping perennial, with a tight central rosette and long shoots with almost hair-like leaves. The flowers are borne on the side-shoots and usually lack petals; May to September. An extremely common weed of damp, bare ground.

Bladder Campion *Silene vulgaris* (1) is a short to tall perennial with pairs of oval, slightly greyish leaves. The flowers are in copious clusters, each 18–20 mm, with deeply cleft petals and a greatly swollen sepal-tube; May to September. Common in dry grassy and bare places in the lowlands. **Sea Campion** *S. maritima* has blue-grey fleshy leaves and many non-flowering shoots, forming a cushion. Sea-shingle and cliffs, rarely on mountains.

Moss Campion *Silene acaulis* is a densely tufted perennial forming bright green, mossy cushions, with slightly woody stems covered in very narrow leaves. The flowers are often produced in great numbers, covering the cushion, and are a beautiful deep rose-pink; June to August. Widespread on cliffs and rock-ledges in mountains, but found at sea-level in N Scotland.

Red Campion *Silene dioica* is a tall, hairy perennial with overwintering rosettes of long-stalked, pointed oval leaves. The flowering stems bear unstalked leaves and abundant bright reddish pink flowers, 18–25 mm; March to November. The seed-pod has short, rolled-back teeth. Pale-flowered forms may be hybrids with White Campion (below). A common and widespread plant of woods and hedges on deep soils; also on sea-cliffs.

White Campion *Silene alba* is very similar to Red Campion (above) and forms fully fertile hybrids with it. It has narrower leaves and larger flowers, 25–30 mm, white and scented in the evening (it is pollinated by moths); May to October. The seed-pods have long, slightly bent teeth. Both species have male and female flowers on separate plants. A common plant of roadsides and bare places, becoming scarce in the W.

Ragged Robin *Lychnis flos-cuculi* is a tall perennial with narrow leaves, shiny but rough to the touch. The flowers are quite unmistakable, 25–40 mm, and with 4-lobed petals, giving the ragged appearance; May to August. Common in damp meadows, marshes and wet woods everywhere.

Soapwort *Saponaria officinalis* is a tall, rather weak-stemmed perennial with large oval leaves and tight clusters of pale pink flowers, 20–30 mm; June to September. The petal tube is distinctly longer than that of the sepals, and the flowers have only 2 styles. A hedgerow plant, nearly always a garden escape, becoming scarce in the N.

BUTTERCUP FAMILY Ranunculaceae have 5 petals and many stamens.

Stinking Hellebore *Helleborus foetidus* is a tall, strong-smelling perennial, with stems that persist through the winter. The leaves have 3–9 long-toothed lobes and all are on the stem, but the uppermost ones are reduced to sheaths. The flowers are bell-shaped, 10–30 mm, with purple-tipped green sepals and no petals; January to May. A scarce native plant of dry, open woods on limestone, more often as a garden escape.

Green Hellebore *Helleborus viridis* is a short perennial appearing in early spring. It has 2 deeply lobed root-leaves and a leafy stem bearing small clusters of drooping flowers, 30–50 mm, with spreading green sepals and no petals; February to April. An uncommon woodland plant, found in deep leaf-mould, but more widespread as a garden escape.

Marsh Marigold *Caltha palustris* is a low to short, stout-stemmed perennial, whose shiny, dark green, kidney-shaped leaves are usually mottled. The flowers have 5 shiny yellow sepals and no petals, and vary greatly in size, 10–50 mm; March to July. The fruits are long and conspicuous, to 15 mm. Widespread and common in wet places.

Meadow Buttercup *Ranunculus acris* (1) is a short or tall perennial with hairy, 3–7-lobed leaves, of which the end-lobe has no stalk. The flowers are 18–25 mm across, with smooth stalks and upright sepals; April to October. A very common and widespread grassland plant. **Bulbous Buttercup** *R. bulbosus* is shorter, has all its leaf-lobes stalked and sepals turned back, and grows in drier places.

Goldilocks Buttercup *Ranunculus auricomus* is a rather variable short perennial with rounded leaves from the stem-base, but deeply and narrowly lobed upper leaves. The flowers are borne singly, 15–25 mm, with up to 5 petals (though it may have none) which are rarely full-formed; April to May. A widespread woodland plant, less common in the N and W.

Creeping Buttercup *Ranunculus repens* is a short, creeping perennial with long, rooting runners. The hairy leaves have 3 lobes, the middle of which is long-stalked. It has large flowers, 20–30 mm, with upright sepals, on long, hairy stalks; May to September. A very common plant of damp grassland and bare places, sometimes a troublesome weed.

38

Greater Spearwort *Ranunculus lingua*, the largest British buttercup, is a tall perennial with long runners and long, toothed, spear-shaped leaves up to 25 cm. The flowers are 20–40 mm, on slightly hairy, furrowed stalks; June to September. Marshes, fens, and lakesides, widespread but declining with its habitat.

Celery-leaved Buttercup *Ranunculus sceleratus* is a short or tall annual with a stout, furrowed stem. The lower leaves are deeply lobed and long-stalked, the upper ones unstalked with 1–3 narrow segments. The flowers are inconspicuous, 5–10 mm; May to September. The fruit-head is conical with 100 or more seeds. It grows on mud, and is common in the S and E, becoming rare in the N.

Lesser Spearwort *Ranunculus flammula* is a very variable low, short or tall perennial, sometimes creeping, sometimes upright, but usually rooting at some leaf-junctions. The leaves are narrow, oval, or occasionally lobed, and the flowers solitary or in sparse clusters, 7–20 mm; June to October. Common in wet, usually grassy places.

Lesser Celandine *Ranunculus ficaria* is a low or short perennial with glossy, dark green, rather fleshy, heart-shaped leaves. The 8–12-petalled flowers arise singly at the ends of the leafy stems, each 20–30 mm; March to May. Common in damp woods and meadows; when growing in deep shade plants may have small bulbils at the base of the leaves.

Common Water Crowfoot *Ranunculus aquatilis* (1) is a very variable aquatic annual or perennial with finely dissected, thread-like submerged leaves and simply-toothed floating leaves. The flowers are white, 10–20 mm; April to September. Common in still or slow water in the lowlands. **River Water Crowfoot** *R. fluitans* has no floating leaves but very long, submerged ones. It grows in fast streams.

Ivy-leaved Crowfoot *Ranunculus hederaceus* (2) is a creeping annual or perennial with all leaves ivy-shaped. The flowers are small, 3–6 mm, with narrow petals no longer than the sepals; April to September. Found on mud and in shallow water scattered throughout Britain. **Round-leaved Crowfoot** *R. omiophyllus* has larger flowers, 8–12 mm. Mainly in the W.

Common Meadow-rue *Thalictrum flavum* (1) is a tall, unbranched perennial with 2 or 3 times pinnate leaves and dense clusters of fragrant flowers, seeming yellow from the conspicuous anthers; June to August. Common in wet meadows and by ditches, becoming rare in the N and W. **Lesser Meadow-rue** *T. minus* has flowers in loose clusters, and is scarce in the S and E.

Columbine *Aquilegia vulgaris* is a tall perennial whose slightly grey–green leaves have 3 3-lobed leaflets. The unmistakable flowers have 5 rich blue–purple petals, each 25–30 mm long, with a long spur, tightly curled at the tip; May to July. Pink- and white-flowered forms occur too. Scattered in woods and scrub on basic soils, and as a garden escape.

Wood Anemone *Anemone nemorosa* is a short perennial with unbranched stems each bearing 3 3- or 5-lobed leaves in a whorl and a single flower, 20–40 mm, with 5–9 (usually 6 or 7) white sepals, pink-tinged on the underside; March to May. The flowers have no petals. Common in deciduous woods except on very acid soils, often carpeting the ground.

Pasque Flower *Pulsatilla vulgaris* is a low or short perennial profusely covered with long, silky hairs. The leaves are twice or thrice pinnate, with very narrow leaflets, and the flowers arise singly on long stalks, erect at first then nodding, bell-shaped, 50–80 mm, and rich purple; April to May. A declining plant of short, dry turf on chalk and limestone.

WATER-LILY FAMILY Nymphaceae root in the mud in still or slow-moving fresh water, and have large, floating leaves and conspicuous flowers.
Yellow Water-lily *Nuphar lutea* is a perennial, rooting in mud in still or slow fresh water. The oval leaves, up to 40 cm, float on the surface, and the almost globular yellow flowers, 40–60 mm, have 5–6 petals and rise out of the water on long stalks; June to September. Widespread, but commonest in the S and E.

White Water-lily *Nymphaea alba* is very similar in habit to Yellow Water-lily (above) but has rounder leaves to 30 cm, and white flowers with 20–25 petals, up to 200 mm across; June to September. A widespread plant of still water in the lowlands.

FUMITORY FAMILY Fumariaceae have 2-lipped flowers and pinnate leaves.

Common Fumitory *Fumaria officinalis* (1) is a scrambling annual with 2 or 3 times pinnate leaves, the segments flat and narrow. The 2-lipped, spurred flowers, 7–8 mm long, are pink and darker-tipped; April to October. A common lowland weed, particularly on sandy soils. **Wall Fumitory** *F. muralis* is more robust and has fewer, larger flowers, 9–12 mm, the lower petal upturned. Much less common.

Yellow Corydalis *Corydalis lutea* is a short, much-branched, leafy perennial with pinnate leaves. The leaflets are trefoil. The 2-lipped yellow flowers, 12–18 mm long, have a short, bent spur and are in spikes of 6–10, opposite the upper leaves; May to September. Introduced on walls in many places.

Climbing Corydalis *Corydalis claviculata* is a delicate climbing annual whose pinnate leaves have trefoil leaflets and end in a long, branched tendril. The 2-lipped creamy flowers, 5–6 mm long, are in sparse spikes opposite the leaves; June to September. Scrambles among vegetation (often bracken) in woods and rocky places on acid soils.

POPPY FAMILY Papaveraceae have pinnate leaves and large, showy, 4-petalled flowers with 2 sepals

Common Poppy *Papaver rhoeas* (2) is a short or tall hairy annual with pinnately cut leaves. The large scarlet flowers, 70–100 mm, have stalks with spreading hairs; June to September. The seed-pods are rounded. A common lowland weed. **Long-headed Poppy** *P. dubium* (3) has smaller, paler flowers on stalks with flattened hairs, and elongated, cylindrical seed-pods.

Yellow-horned Poppy *Glaucium flavum* is a tall, grey–blue perennial with large, pinnately lobed basal leaves and wavy-edged upper leaves that clasp the stem. The flowers are large, 60–90 mm, on short, hairless stalks; June to September. The seed-pods are very long, up to 300 mm. On shingle by the sea from S Scotland southwards.

Greater Celandine *Chelidonium majus* (4) is a tall perennial whose stems break easily and exude orange juice. The leaves are slightly grey and coarsely pinnately lobed. Flowers are 20–25 mm; April to September. Hedges and waste places, usually near houses. **Welsh Poppy** *Meconopsis cambrica* is shorter, more compact, hairier, and has flowers 50–75 mm. Native in Wales and SW England, naturalised elsewhere.

seed-pod
of 3

41

CABBAGE FAMILY Cruciferae have 4 petals and 4 sepals, making a cross, and distinctive long or rounded seed-pods.

Wild Turnip *Brassica rapa* (1) is a tall annual or biennial with large, deeply-lobed, bristly lower leaves, and small, narrow upper leaves with lobes clasping the stem. The flowers, 20 mm, hide the unopened buds; April to August. Pods are long. Disturbed ground, often on river-banks. **Black Mustard** *B. nigra* is greyish, with stalked stem-leaves and pods flat against the stem.

Treacle Mustard *Erysimum cheiranthoides* is a short or tall annual whose square stems are covered with flattened, branched hairs. It has narrow, slightly toothed leaves and small flowers, 6 mm across; June to September. The 4-angled pods are up to 30 mm long. Disturbed ground; most abundant in E Anglia, and rare in the W and N.

Great Yellowcress *Rorippa amphibia* is a tall, hollow-stemmed annual spreading by runners. The narrow oval leaves may be toothed or more deeply lobed, and the flowers are small, 6 mm; June to August. It has oblong pods on long, stiff, spreading stalks. In and beside still and slow-flowing water in central England.

Creeping Yellowcress *Rorippa sylvestris* (2) is a short, weak-stemmed perennial spreading by runners. The pinnate leaves have deeply-lobed segments and the flowers, 5 mm, have petals longer than the sepals; June to September. The pods are cylindrical, up to 18 mm long, longer than their stalks. Damp, bare places, except in the extreme N and W. **Marsh Yellowcress** *R. palustris* is more erect, has 3 mm flowers with sepals as long as the petals, and small, oblong pods.

Hedge Mustard *Sisymbrium officinale* (3) is a tall, stiffly erect annual with pinnately lobed leaves. It has pale yellow flowers, 3 mm across, from May to September and short, cylindrical pods, 6–20 mm, closely pressed to the stem. Very common on disturbed ground. **Eastern Rocket** *S. orientale* has flowers 7 mm across and long, spreading pods; rare in Scotland. **Tall Rocket** *S. altissimum* has 10 mm flowers, deeply cut leaves, and long, spreading pods.

Charlock *Sinapis arvensis* (4) is a tall, roughly hairy annual with large, coarsely and unevenly-lobed lower leaves. The upper leaves are unstalked but, unlike Wild Turnip (above), not clasping the stem. The flowers are 15–20 mm, from April to October, and the pods long and beaded. Common in disturbed ground. **White Mustard** *S. alba* has all its leaves pinnately lobed.

Annual Wall Rocket *Diplotaxis muralis* is a short, hairless annual with a rosette of pinnately lobed leaves and one or several almost leafless and often branched flowering stems. The deep yellow flowers are 10–15 mm across; June to September. The pods are long and thin, held upright on shorter, spreading stalks. Dry, bare places and walls, becoming rare in the N.

Common Wintercress *Barbarea vulgaris* is a tall, shiny green perennial with unevenly pinnate leaves in a basal rosette and almost entire stem-leaves. It has bright yellow flowers, 7–9 mm, from May to August, and long, 4-angled pods held stiffly upright. Common in hedges, on river-banks, and in damp, disturbed ground, except in upland areas.

Watercress *Nasturtium officinale* is a creeping, hollow-stemmed perennial with pinnate leaves. The flowers are white, 4–6 mm, from May to October, and the spreading, oblong or slightly curved pods are up to 18 mm long and beaded. Common in shallow water and springs and on mud, except in the mountains.

Hairy Bittercress *Cardamine hirsuta* (1) is a low, slightly hairy annual with a neat rosette of pinnate leaves. It has inconspicuous white flowers, 2–4 mm, with 4 stamens and often concealed by the long, upright pods. It flowers nearly all year round. Very common on bare ground. **Wavy Bittercress** *C. flexuosa* is often larger, has zigzag stems, leaves with up to 15 leaflets, flowers with 6 stamens from April to September, and pods not overtopping the stem. A plant of damp, shady places.

Cuckoo Flower *Cardamine pratensis* (2) is a short perennial with a rosette of pinnate leaves with oval or rounded leaflets, and stem-leaves which have very narrow leaflets. The flowers are lilac or white, 12–20 mm, with yellow anthers; April to June. The long-stalked, narrow pods, up to 40 mm, point upwards. Very common in damp grassland. **Large Bittercress** *C. amara* has broader leaflets on the stem-leaves and white flowers with violet anthers.

Wild Radish *Raphanus raphanistrum* is a tall, roughly hairy annual with coarsely pinnately lobed lower leaves and blunt-toothed stem-leaves. The flowers may be yellow or white with lilac veins, 25–30 mm; May to September. The pods are long, beaded and fragile. A common plant in disturbed ground in light soils, except in hilly areas.

43

Wild Candytuft *Iberis amara* is a short, stiffly erect and much branched perennial with rather narrow, usually toothed leaves, scattered up the stem irregularly. The white or mauve flowers, 6–8 mm, have 2 long and 2 short petals and are crowded in flat-topped heads, elongating in fruit; May to September. Pods are small, round, and long-stalked. Native on dry, bare, chalky soils in S England, but a more widespread garden escape.

Sweet Alison *Lobularia maritima* is a low, hairy perennial with narrow, greyish leaves. The white flowers, 5–6 mm, are fragrant; June to October. The pods are small and oval, on spreading stalks much longer than themselves. This is the familiar garden edging plant which often escapes and is frequently naturalised.

Shepherd's Purse *Capsella bursa-pastoris* is an extremely variable low or short annual, sometimes overwintering. The leaves are rather narrow and may be untoothed, lobed, or almost pinnate, but are usually mainly in a basal rosette. It flowers all year round, but the flowers are inconspicuous, 2–3 mm, and the pods form an inverted, notched triangle from which its name derives. One of the commonest weeds.

Common Whitlow-grass *Erophila verna* is a variable low annual with a neat rosette of narrow, toothed leaves and several leafless stems. The white flowers, 3–6 mm, have deeply notched petals; March to May. The flattened pods are elliptical or rounded, on long stalks. A common and often over-looked plant of dry, bare, often sandy or rocky ground.

Field Pennycress *Thlaspi arvense* is a short annual with a strong, rather unpleasant smell, and no basal rosette of leaves. It has pointed, toothed leaves clasping the stem, and long spikes of small white flowers, 4–6 mm; May to September. The rounded, flattened pods have a deep notch and broad, clear margins. A common arable weed in lowland areas, often seen as persistent dead stems with pods in winter.

Field Pepperwort *Lepidium campestre* is a short, greyish, densely hairy annual or biennial with branching stems. Its lower leaves are stalked, the upper ones triangular and clasping the stem. The flowers are 2–3 mm in a crowded spike; May to August. The notched pods are oblong and white-spotted. Widespread in dry, rather bare places, becoming rare in the N.

Scurvy Grass *Cochlearia officinalis* (1) is a variable low or short perennial, often purple-stemmed. It has dark green, fleshy, heart-shaped leaves, the lower ones long-stalked, the upper toothed and clasping the stem. The flowers, 8–10 mm, from April to August, ripen to almost spherical pods. In salt-marshes, on cliffs, and in mountains on wet rocks. **Early Scurvy-grass** *C. danica* is a smaller annual with all leaves stalked, lilac flowers, and oval pods. Sandy seashores.

Thale Cress *Arabidopsis thaliana* is a short annual with a rosette of elliptical, toothed, roughly hairy leaves. The stems are often branched and bear few, narrow, unstalked leaves. The flowers are 3 mm across from March to October, and the long, almost cylindrical pods are held upright. A common weed of dry, sandy, often acid soils, as well as walls and gravel paths.

Horseradish *Armoracia rusticana* is a massive, tall perennial with very large, long-stalked root-leaves which have wavy, toothed edges. The stem-leaves are narrow, each giving rise to a long spike of white flowers, 8–9 mm; May to August. The pods are long-stalked and spherical. Cultivated for its edible root, and widely naturalised in grassy and waste places.

Hoary Cress *Cardaria draba* is a short or tall branching perennial with narrow, pointed, greyish leaves clasping the stem. The white flowers, 5–6 mm, are in flat-topped clusters, so that at a distance the plant looks like an umbellifer; May to June. The pods are oval or kidney-shaped. On roadsides and as an arable weed, commonest in the SE and rare in Scotland, but spreading.

Swinecress *Coronopus squamatus* is a low, creeping annual or biennial whose greyish, twice pinnate leaves have narrow lobes. The tiny white flowers, 2–3 mm, are in tight clusters opposite the leaves; June to September. The rounded, long-stalked pods are warty and 2-lobed. Typically grows in bare, firm, trodden ground, such as paths and gateways, commonest in the SE.

Lesser Swinecress *Coronopus didymus* is very similar to Swinecress (above) but has more finely divided, feathery leaves, with very narrow segments. The flowers are even smaller, 1–2 mm, and sometimes have no petals; June to September. The pods are longer-stalked and deeply notched. Bare places and disturbed ground, commonest in the S and SW and rare N of the Humber.

45

Garlic Mustard *Alliaria petiolata* is a tall perennial with long-stalked, toothed, heart-shaped leaves on the stem, which smell somewhat like garlic when crushed. The flowers are white, 6 mm, in almost leafless spikes; April to June and sparsely till August. The 4-angled pods are up to 60 mm long, curving upwards then vertical. A very common hedgerow plant, also found in open woods; becoming rare in the NW.

Sea Rocket *Cakile maritima* is a short annual with hairless, greyish, fleshy stems and leaves. The lower leaves are pinnately lobed. The flowers are usually lilac, but may be darker or white, up to 20 mm, in dense spikes; June to September. The oval pods are jointed near the base. Common on sandy shores around the high-tide mark.

Sea Kale *Crambe maritima* is a massive but not tall, greyish perennial whose stems are woody at the base. The oval lower leaves are up to 30 cm long, long-stalked, fleshy and wavy-edged, but those on the stem are narrower. The flowers are 10–15 mm across, in rather flat heads; June to August. The erect, oval pods have spreading stalks. On sandy and shingly seashores from S Scotland southwards.

Dame's Violet *Hesperis matronalis* is a tall, branching, hairy perennial with narrow, finely-toothed leaves on short stalks. The very fragrant flowers are 15–20 mm across, and white or violet; May to August. The upright cylindrical pods are 80–90 mm long. A widespread garden escape by roads and paths.

MIGNONETTE FAMILY   Resedaceae
Weld *Reseda luteola* is a tall biennial with a rosette of narrow, unstalked leaves, and in the second year an unbranched, leafy flowering stem. Stem-leaves are similar, untoothed but wavy-edged. The yellow–green flowers, 4–5 mm, have 4 sepals and petals and are in long, leafless spikes; June to September. Common in the SE on disturbed, usually lime-rich ground, becoming rarer in the N and W.

Wild Mignonette *Reseda lutea* is a tall, branching biennial or perennial, though usually shorter than Weld (above). The stem is covered with many small pinnate leaves, and the 6-petalled flowers, 6 mm across, are in a shorter, stouter spike; June to September. It grows in similar places to Weld, but with a more southern distribution.

SUNDEW FAMILY Droseraceae are insectivorous plants.

Common Sundew *Drosera rotundifolia* (1) is a low perennial with a neat, flat rosette of rounded leaves covered with long, red, sticky hairs which curl inwards to trap and digest insects. It has a spike of white flowers, 5 mm, on an upright stalk; June to August. In wet, peaty places, and so commonest in N and W. **Great Sundew** *D. anglica* has larger, narrow leaves; mainly in the N. **Oblong-leaved Sundew** *D. intermedia* is like a small Great Sundew with the flower-stalk starting beneath the rosette.

STONECROP FAMILY Crassulaceae are fleshy perennials with 3-petalled, star-like flowers.

Navelwort *Umbilicus rupestris* is a short perennial with round, fleshy leaves, sunken in the middle like a navel, above the leaf-stalk. It produces a long, leafless spike of greenish or creamy, bell-shaped flowers, 8–10 mm long, from June to August. In cracks in acid rocks and on stony banks, mainly in Wales and the SW.

Roseroot *Rhodiola rosea* is a short, stocky, greyish perennial, often red-tinged. The stout stem is almost hidden by the thick, fleshy, oval leaves and ends in a tight, rounded head of dull yellow flowers, 4–6 mm, turning orange in fruit. Male and female flowers on separate plants; May to July. On mountain rocks and on sea-cliffs in N Scotland.

Orpine *Sedum telephium* is a short, slightly bluish perennial with little knots of reddish stems, bearing large, fleshy, oblong leaves. The flowers are dull purple or sometimes greenish yellow, 9–12 mm, in tight heads; July to August. Widespread in woods and hedges, occasionally also on rocks, except in the Midlands and NW Scotland.

English Stonecrop *Sedum anglicum* (2) is a deep red, creeping perennial forming mats of densely leafy stems. The fleshy, cylindrical leaves are 3–5 mm long, and the white flowers, 10–12 mm, are in a sparse, leafy head; June to September. On rocks and walls, and rarely sand, mainly in the W. **White Stonecrop** *S. album* has longer green leaves and branching heads of smaller flowers. Less common but more widely distributed.

Biting Stonecrop *Sedum acre* is a creeping green perennial, hot to the taste. The short, fleshy, cylindrical leaves are pressed against the stem. It has bright yellow flowers, 10–12 mm; May to July. Common on bare ground.

47

GRASS OF PARNASSUS FAMILY
Parnassiaceae

Grass of Parnassus *Parnassia palustris* is a short perennial with a loose rosette of long-stalked, heart-shaped leaves. The flowering stems each bear 1 unstalked leaf and 1 large white flower, 15–30 mm, with prominent greenish veins; June to September. Wet, often peaty places; also in dune-slacks. Common in W Scotland and NW England, rarer S to the Midlands.

SAXIFRAGE FAMILY
Saxifragaceae are small perennials with 5 petals and 10 stamens.

Meadow Saxifrage *Saxifraga granulata* is a short, sparsely hairy perennial, producing bulbils in autumn at the base of the leaves. It has kidney-shaped leaves with rounded lobes, and often almost leafless flower-stems, topped by loose clusters of unspotted white flowers, 20–30 mm, on stickily hairy stalks; April to June. A declining meadow species, scattered through most of the country.

Starry Saxifrage *Saxifraga stellaris* (1) is a low perennial with a tight rosette of thick, almost unstalked leaves. It has open clusters of white flowers, 10–15 mm, which appear red-dotted from the anthers; June to August. The commonest white-flowered mountain saxifrage, on wet rocks and by streams. **Mossy Saxifrage** *S. hypnoides* forms a tight cushion, with masses of narrow 3-lobed leaves. On hills rather than mountains only.

Yellow Saxifrage *Saxifraga aizoides* is a low perennial with masses of very leafy stems, half-creeping and then turning upwards, forming mats. It has narrow, slightly fleshy leaves and yellow flowers, 10–15 mm, with the green sepals visible between the petals; June to September. Wet places in mountains of N England and Scotland.

Purple Saxifrage *Saxifraga oppositifolia* is a creeping perennial with tough, trailing stems clothed with tiny bluish leaves in opposite pairs. The flowering branches are less leafy and bear single purple flowers, 10–15 mm, in March and April and less abundantly in July and August. On mountain rocks from the Brecon Beacons northwards.

Opposite-leaved Golden Saxifrage *Chrysosplenium oppositifolium* is a creeping, patch-forming perennial with square stems bearing opposite pairs of rounded, toothed leaves. The tiny flowers, 3–4 mm, have no petals but the yellow anthers are conspicuous; March to July. In wet places, often in springs; common in the N and W but rare in or absent from E England.

ROSE FAMILY Rosaceae have 5-petalled flowers, mostly showy, and pinnate or palmate leaves, except for woody species (pp 22–4).

Meadowsweet *Filipendula ulmaria* (1) is a tall perennial with dark green, pinnate leaves, often silvery on the underside. The oval leaflets are toothed and the terminal one is 3- or 5-lobed. The creamy flowers, 4–6 mm, are in dense, branching, arching spikes; June to September. Very common in wet places. **Dropwort** *F. vulgaris* is smaller, with many more pairs of leaflets. Dry grassland on chalk or limestone; commonest in the S.

Agrimony *Agrimonia eupatoria* is a short or tall perennial with a narrow, upright spike bearing pinnate leaves with coarsely toothed leaflets and crowded yellow flowers, 5–8 mm; June to August. The grooved fruit has a ring of hooked hairs around it, which cause it to cling to fur, clothing, and the like. A common grassland plant, particularly on roadsides, except in N Scotland.

Great Burnet *Sanguisorba officinalis* is a tall, well-branched but slender perennial, with dark green pinnate leaves, each with 3–7 pairs of oval leaflets, 20–40 mm long. The tiny flowers are densely packed in oblong, deep crimson heads, 10–20 mm long; June to September. Common in damp grassland and fens in N England and the Midlands, scarcer elsewhere in England and Wales and rare in Scotland.

Salad Burnet *Sanguisorba minor* is a short perennial whose bluish pinnate leaves have up to 12 pairs of rounded leaflets. The flowers are in round heads, 6–12 mm across, with red styles in the topmost flowers and yellow stamens in the lower ones; May to September. In chalk and limestone grassland, north to S Scotland.

Lady's Mantle *Alchemilla vulgaris* is an aggregate name for a very complex group of species, all of which are short perennials with large, rounded, palmately lobed leaves. The tiny, yellow–green flowers, 2–3 mm, are in small clusters in a branched head; May to September. Grassland, usually in upland areas. The commonest species are *A. glabra*, with no hairs, and *A. xanthochlora*, with long, silky hairs.

Parsley Piert *Aphanes arvensis* is an insignificant prostrate annual with very small, 3-lobed leaves and cup-like stipules at the base of the leaves. It has minute green flowers, 1–2 mm, in clusters opposite the leaves; April to October. Bare ground, rare in Scotland.

Dog Rose *Rosa canina* (1) is a shrub with arching stems, reaching 3 m, and armed with curved thorns. The hairless pinnate leaves have 2–3 pairs of toothed leaflets, and the pale pink or white flowers, 45–50 mm, have a flat disc in the centre, comprising the styles; June to July. The fruit is the familiar red rose-hip. Common in woods, hedges, and wooded fens. **Downy Rose** *R. tomentosa* is a northern species with downy leaves and smaller, deep pink flowers.

Field Rose *Rosa arvensis* is a rather weak-stemmed shrub often forming dense mounds, less than 1 m high. It has green or purplish stems with curved thorns, pinnate leaves with 2–3 pairs of leaflets, and white flowers, 30–50 mm, with a prominent central column of styles; July to August. Common in woods and hedges in S England, becoming rare in the N and absent from Scotland.

Burnet Rose *Rosa pimpinellifolia* is a short shrub, rarely more than 50 cm high, whose stems have stiff bristles and straight thorns. The leaves have 3–5 pairs of rounded, sometimes bluish leaflets, and the creamy white flowers, 20–40 mm, appear from May to July. The ripe hips are spherical and purplish black. On sand-dunes all around the coast, and in limestone grassland in the N and W.

Bramble *Rubus fruticosus* is an extremely variable collection of hundreds of similar species of scrambling shrubs with curved thorns. The arching, biennial stems root where they touch the ground and bear prickly, pinnate leaves. The white or pink flowers, 20–30 mm, appear from May to November, and the purplish black fruits (blackberries) from August. Very common in woods and hedges and on bare ground.

Raspberry *Rubus idaeus* is a very tall perennial whose woody biennial stems have weak prickles. The leaves have 3–5 leaflets which are greyish underneath, and the flowers, 15–20 mm, have narrow petals; May to August. The fruit is the familiar raspberry, though smaller than in cultivated forms. Woods, scrub, and heaths, commonest in the N.

Wild Strawberry *Fragaria vesca* is a low perennial with long, creeping runners, rooting and carrying clusters of bright green trefoil leaves, hairy on the undersides. It has white flowers, 12–18 mm, from April to July, and small, very sweet strawberries. Common in dry, grassy places and in woods.

Barren Strawberry *Potentilla sterilis* is like a small, bluish Wild Strawberry (opposite) except for its quite different, dry fruit. It has hairy stems and blunt-toothed leaves. The flowers are 10–15 mm across, with notched petals; February to May. Common in dry places and open woods, becoming scarce in the far N.

Marsh Cinquefoil *Potentilla palustris* is a short perennial with sharply-toothed, pinnate leaves near the base of the stem, and 3- or 5-lobed leaves at the top. The star-shaped flowers, 20–30 mm, have conspicuous, spreading, brownish purple sepals, and much smaller, purple petals; May to July. Common in moors, wet meadows, and in shallow water in the N and W, usually on acid soils; rare in the SE.

Tormentil *Potentilla erecta* (1) is a low, creeping perennial, often forming tight clumps. Its leaves are apparently 5-lobed (the lowest lobes are actually stipules), with narrow-toothed lobes. It produces several loose heads of 4-petalled flowers, 7–11 mm; May to September. Very common on acid grassland, heaths, and moors. **Creeping Cinquefoil** *P. reptans* has long rooting stems, 5–7-lobed leaves and single flowers, 17–25 mm. Less common in Scotland.

Silverweed *Potentilla anserina* is a low, creeping perennial with silvery, pinnate leaves which have up to 12 pairs of leaflets and diminutive ones in between them. The yellow flowers, 15–20 mm, are borne singly on long, upright stalks, but are often almost absent; May to August. Common in damp grassy and bare places.

Herb Bennet *Geum urbanum* is a short, rarely tall perennial with downy stems and pinnate leaves, with a large, lobed leaflet at the end. It has spreading flowers, 8–15 mm, from May to September, with a long, hooked, hairy style, which persists on the fruits. Common in damp woods and hedges on deep soils. It hybridises with Water Avens (below) wherever they grow together, producing a confusing range of intermediates.

Water Avens *Geum rivale* is a short, downy perennial with pinnate lower leaves and trefoil stem-leaves. It has nodding, bell-shaped flowers, up to 15 mm long, with pointed, dull purple sepals, and slightly notched, apricot-pink petals; April to September. The fruits have hooked styles, as in Herb Bennet. Common in the N in damp, often wet and shady places; rare in the S and SE.

PEA FAMILY Leguminosae have 5-petalled flowers – a broad 'standard' at the top, 2 'wings' at the side, and 2 joined in a 'keel' concealing the stamens. Leaves mostly pinnate or trefoil.

Gorse *Ulex europaeus* (1) is a spiny, evergreen shrub, usually 1–2 m high. It has green stems and the leaves are furrowed green spines. The fragrant flowers, 15–20 mm, with wings longer than the keel, appear all year, but mainly February to May. Very common on light, usually acid soils. **Western Gorse** *U. gallii* forms tighter bushes, with almost smooth spines. Equally common in Wales and the SW, scarce elsewhere.

Broom *Cytisus scoparius* is a spineless, deciduous shrub, up to 2 m high, with ridged green twigs, and short-stalked, trefoil leaves. The flowers are up to 20 mm, on slender stalks; April to June. The pods are long and black, with brown hairs, opening with an explosive crack in dry weather. Common on heaths and open woods on sandy soils.

Dyer's Greenweed *Genista tinctoria* is a spineless shrub up to 70 cm high, with narrow, pointed leaves and slim, yellow flowers, 15 mm, in leafy, stalked spikes; June to August. The standard is as long as the keel. Locally common in unimproved grassland and open woods north to S Scotland.

Petty Whin *Genista anglica* is a small, spiny, slender shrub, rarely more than 50 cm high, with narrow oval, rather bluish leaves. It has slightly flattened, yellow flowers, 8–10 mm, with the standard shorter than the keel; April to June. Scattered throughout the country on heaths and moors.

Purple Milk-vetch *Astragalus danicus* is a low perennial, clothed in short, soft, white hairs. It has pinnate leaves with up to 10 pairs of oval leaflets, and long-stalked, tight heads of upright, blue–purple flowers, 15 mm long; May to July. The ripe pods are covered with white hairs. Rather scarce in short grassland on chalk and limestone, and in dunes; mainly in the E.

Wild Liquorice *Astragalus glycyphyllos* is a tall, straggling, thick-stemmed perennial, whose pinnate leaves have oblong leaflets. The creamy white, often green-tinged flowers, 10–15 mm, are in loose heads, on stalks shorter than the leaves; June to August. A rather scarce plant of rough grassland and scrub; widespread but mainly in central England.

Tufted Vetch *Vicia cracca* (1) is a scrambling perennial with stems up to 2 m long, bearing long, pinnate leaves with up to 12 pairs of leaflets ending in a branched tendril. It has dense, 1-sided spikes of up to 40 bluish violet flowers, 10–12 mm long; June to August. The pods are brown when ripe, 10–20 mm long. Common in tall grassland and scrub. **Wood Vetch** *V. sylvatica* (2) has broader leaflets and larger, very pale violet flowers with purple veins. Woods, scattered.

Bush Vetch *Vicia sepium* is a tall, clambering perennial which has pinnate leaves with 5–9 pairs of oval leaflets, ending in a branched tendril. The dull purple and blue flowers, 12–15 mm, are in short spikes of 2–6 flowers; April to October. The ripe pods are black. Very common in rough, grassy places and hedges.

Common Vetch *Vicia sativa* is a very variable annual, either short or tall and scrambling. The pinnate leaves have 3–8 pairs of oval, or sometimes very narrow leaflets, and branched or unbranched tendrils. The clear purple–pink flowers, 10–30 mm long, are typically in pairs, though occasionally single or rarely in fours; April to September. The pods are black. Common in bare and grassy places and hedges.

Hairy Tare *Vicia hirsuta* (3) is, somewhat paradoxically, a hairless, weak-stemmed, scrambling annual. It has pinnate leaves with 4–8 pairs of very narrow leaflets, and usually branched tendrils, and slender spikes of 1–9 pale lilac–blue flowers, 4–5 mm long; May to September. The ripe pod is black and hairy. Common in grassy places, scarcer in Wales and Scotland. **Smooth Tare** *V. tetrasperma* has pairs of deep lilac flowers with 3 long and 2 short sepal-teeth, and smooth pods.

Bitter Vetchling *Lathyrus montanus* is a short perennial with a winged stem and pinnate leaves, with 2–4 pairs of oblong leaves, and no tendrils. It has short spikes of 2–6 flowers, 10–15 mm, opening a distinctive rich reddish purple and fading to purplish blue; April to July. The ripe pods are red–brown. Common in open woods, heaths and rough grassland in hilly areas, but absent from much of E England.

Meadow Vetchling *Lathyrus pratensis* is a short or tall, downy perennial, scrambling over other plants. It has angled, but unwinged stems, and the leaves have tendrils and only 1–2 pairs of narrow, pointed leaflets. The flowers are in clusters of 4–12, each 15–18 mm long; May to August. Ripe pods are black. Very common in grassy places.

Sainfoin *Onobrychis viciifolia* is a short, downy perennial with pinnate leaves, each with 6–12 pairs of narrow, oval leaflets, and ending in a leaflet. The red-veined, bright pink flowers, 10–12 mm, are in long, dense spikes; June to August. The ripe pods are warty, 6–8 mm long. Native in dry chalk and limestone grassland in S and E England, but cultivated and naturalised elsewhere.

Rest-harrow *Ononis repens* (1) is a half-creeping, hairy shrub, sometimes armed with soft spines. It has trefoil leaves with oval leaflets and pink flowers, 10–15 mm, with the wings as long as the keel; July to September. Common in dry grassland in the S, becoming restricted to coastal regions in Scotland. **Spiny Rest-harrow** *O. spinosa* (2) has spiny stems with 2 lines of hairs; the wings of the reddish pink flowers are shorter than the keel. More southern.

Ribbed Melilot *Melilotus officinalis* (3) is a tall biennial whose trefoil leaves have toothed, oblong leaflets. The flowers, 5–6 mm, are in long spikes; June to September. Ripe pod is brown. Naturalised in waste places, but rare in Scotland. **Tall Melilot** *M. altissima* has a longer keel to the flower, and downy black pods. **White Melilot** *M. alba* (4) has white flowers.

Kidney Vetch *Anthyllis vulneraria* is a short perennial with long, silky hairs on its stems, its narrow, pinnate leaves, and the sepal-tubes. It has yellow, orange, red, purple or mixed flowers, 12–15 mm long, in a dense, rounded head; May to September. Widespread and often abundant in dry grassland, particularly near the sea.

Horseshoe Vetch *Hippocrepis comosa* is a low, spreading perennial with pinnate leaves which have 4–5 pairs of narrow leaflets (though occasionally many more) and end in a leaflet. The yellow flowers, 6–10 mm, are in circular 5–8-flowered heads; May to July. The ripe pods are crinkled into horseshoe-like segments. Widespread in dry chalk and limestone grassland in S England, rare in N England.

Birdsfoot Trefoil *Lotus corniculatus* (5) is a low perennial whose leaves have 5 leaflets, the lowest 2 of which are bent back, giving a trefoil appearance. It has heads of 2–6 yellow or orange flowers, 10–15 mm; May to September. Very common in dry, bare and grassy places. **Greater Birdsfoot Trefoil** *L. uliginosus* is larger and stoloniferous, with greyish, often downy leaves and spreading sepals when in bud. Damp places.

variations in colour

Lucerne *Medicago sativa* (1) is a short or tall perennial whose trefoil leaves have toothed leaflets, widest at the tip. It has tight spikes of purple flowers, 7–8 mm; June to October. A subspecies, *falcata* (2), has yellow flowers, and crosses may be green or almost black. Subspecies *sativa* is widely planted and often naturalised, but *falcata* is a rare native of E Anglia.

Black Medick *Medicago lupulina* (3) is a low, downy annual which has trefoil leaves ending in a minute point in the central notch. The yellow flowers, 3–8 mm, are in dense heads of 10–50; April to October. Ripe pods are coiled and black. A common grassland plant, becoming scarce in N Scotland. **Lesser Trefoil** *Trifolium dubium* has no point at the leaf-tip, fewer flowers (up to 15 in a head), and straight pods, hidden by dead flowers.

Spotted Medick *Medicago arabica* is a low, prostrate perennial with trefoil leaves, each leaflet dark-spotted. The leaflets are larger than in Black Medick (above) and the leaves are on long stalks, almost hiding the clusters of 1–4 yellow flowers, 4–7 mm; April to September. It has coiled, spiny pods. In dry grassy and bare places in S England.

Hop Trefoil *Trifolium campestre* is a short, hairy annual with long-stalked, trefoil leaves. It has dense, round heads of flattened, yellow flowers, 10–15 mm, May to September. The dead flowers turn pale brown and hide the pods. Dry grassland, becoming scarce in the extreme NW.

Red Clover *Trifolium pratense* (4) is a short, sometimes tall, hairy, upright perennial with trefoil leaves. The leaflets usually bear a pale, crescent-shaped mark. It has large, oval heads of pinkish purple flowers, 15–18 mm; May to October. A very common grassland plant, widely sown. **Zigzag Clover** *T. medium* has narrow stipules around the leaf-stalks, and long-stalked flower-heads.

Strawberry Clover *Trifolium fragiferum* is a creeping perennial with rooting stems, and trefoil leaves which have toothed leaflets. It has small, round, long-stalked heads of pink flowers, June to September. In fruit they swell to look slightly like brownish pink strawberries. In grassland, mainly in S and E England.

55

White Clover *Trifolium repens* (1) is a creeping perennial with rooting stems bearing trefoil leaves, variously marked, on long or short stalks. The creamy white or greenish white flowers are in a rounded head, 15–30 mm across; May to October. An extremely common grassland plant, widely sown in leys. **Alsike Clover** *T. hybridum* is short and upright with pinkish flowers. Widely cultivated.

Haresfoot Clover *Trifolium arvense* (2) is a short, hairy, upright annual with narrow leaflets. It has long, cylindrical, woolly-looking flower-heads of small, pink flowers, 3–4 mm, with long sepal-teeth; June to September. Common on light acid soils in the SE, becoming rare in the N. **Crimson Clover** *T. incarnatum* (3) is altogether larger, with broader leaflets and crimson flowers. Cultivated in the S.

WOOD-SORREL FAMILY
Oxalidaceae

Wood Sorrel *Oxalis acetosella* has leaves like a clover, but is not related. It is a low, creeping perennial and, like the clovers, its trefoil leaves close downwards at night. Its white flowers are lilac-veined and cup-shaped, 15–20 mm, borne singly on slender stalks at about the level of the leaves; April to June. Very common in woods on deep soils, as well as on mountain rocks.

Yellow Oxalis *Oxalis corniculata* is a delicate, low annual or perennial with hairy, rooting stems. It has trefoil leaves, but the leaflets are 2-lobed, and the yellow flowers are in small clusters, on stalks that bend back in fruit; May to October. Most frequent as a garden weed, particularly in the SW, but as a casual north to S Scotland.

FLAX FAMILY   Linaceae

Purging Flax *Linum catharticum* is a dainty, low annual with slender stems clothed with opposite pairs of oblong leaves. The stems end in loose, branching clusters of white flowers, 8–10 mm; May to September. Very widespread in short, unshaded grassland, both wet and dry. Can be confused with sandworts (p. 35), which have 3 styles.

GERANIUM   FAMILY   Geraniaceae have deeply lobed leaves and 3-petalled, often conspicuous flowers. Fruits have a long beak.

Common Storksbill *Erodium cicutarium* is a rather stickily hairy, short annual. It has feathery, twice pinnate leaves and purplish pink flowers, 10–15 mm; May to August. The petals fall soon after the flowers open. The fruit has a long, twisted beak. Very common in the SE in dry bare and grassy places, but further N mainly near the sea.

Meadow Cranesbill *Geranium pratense* (1) is a tall perennial with large violet-blue flowers, on stalks which bend down after flowering before becoming erect again as the long-beaked fruit ripens. The lower leaves are deeply 5- or 7-lobed. Flowers June to September, typically on roadsides, mainly in limestone areas. **Wood Cranesbill** *G. sylvaticum* has smaller, purpler flowers, finished by late July, and less deeply divided leaves. Typical of hay meadows in upland areas.

Bloody Cranesbill *Geranium sanguineum* is a short, often neatly domed perennial of dry limestone soils. Its leaves vary greatly, from almost entire to deeply dissected, though usually the latter. Flowers are a startling bright purplish red, from June to August. Scattered through Britain except the Midlands, the SE and NW Scotland.

Herb Robert *Geranium robertianum* is a rather untidy annual or biennial with 5-lobed lower leaves which often turn red or orange. Upper leaves are 3-lobed and all are deeply dissected. The pale or bright pink flowers, 20 mm, have unnotched petals and orange anthers; April to October. Common in woods, on rocks and by the sea, everywhere except the extreme N.

Dovesfoot Cranesbill *Geranium molle* is a low, often almost creeping, very hairy annual. It has rounded leaves, those at the base of the stem dissected to less than half-way. The sometimes purplish pink flowers, 8–12 mm, have deeply notched petals; April to September. A common plant of dry places and short grassland throughout Britain, but rare in upland areas.

Cut-leaved Cranesbill *Geranium dissectum* is a short, untidy annual with deeply dissected, finger-like leaves. Flowers are purplish pink, 8–10 mm, with notched petals; May to September. Common in grassy, often disturbed places throughout Britain, except in mountains. **Long-stalked Cranesbill** *G. columbinum* has larger flowers on longer stalks, unnotched petals, and more feathery leaves. Dry places, mainly in the S.

Hedgerow Cranesbill *Geranium pyrenaicum* (2) is a tall perennial with rounded leaves cut into 5–9 shallow lobes. Flowers are usually purple, 15–20 mm, in pairs and with deeply notched petals; June to September. Fruit-stalks are long and curve down and up again. Commonest in hedges and mainly in the S and E, but scattered N to the Moray Firth; prefers limestone soils.

SPURGE FAMILY Euphorbiaceae have milky juice in their stems, simple leaves, and curious green flowers (see Wood Spurge below).

Wood Spurge *Euphorbia amygdaloides* (1) is a tall, downy perennial with dark green, unstalked leaves, some in a whorl. The yellow flowers are in loose clusters; April to May. All spurges have flowers without sepals and petals; one female and several male flowers are surrounded by a cup-shaped structure with conspicuous glands, and that in turn by bracts. Wood Spurge has horned glands and fused bracts. Common in woods in S England and Wales.

Sea Spurge *Euphorbia paralias* (2) is a short, greyish perennial, with thick, fleshy, leathery leaves, densely arrayed up the stem. The flowers (see Wood Spurge above) have short-horned glands and rounded bracts; June to October. On sand-dunes, but absent from most of Scotland and NE England S to the Wash. Hard to distinguish from another sand-dune plant, **Portland Spurge** *E. portlandica* (3), which has less leathery leaves with a prominent midrib beneath, and long-horned glands.

Sun Spurge *Euphorbia helioscopia* (4) is a short, unbranched annual with blunt-tipped leaves, widest near the tip, and the upper ones minutely toothed. The flowers (see Wood Spurge above) have leaf-like bracts and simple glands; April onwards. A common weed of cultivated ground throughout lowland Britain. **Broad-leaved Spurge** *E. platyphyllos* (5) has blunt-tipped, heart-shaped leaves. SE England.

Dwarf Spurge *Euphorbia exigua* (6) is a small, greyish annual with very narrow, unstalked leaves. The flowers (see Wood Spurge above) have horned glands and tri-angular bracts. An arable weed, common S but rare N of a line from the Humber to the Wash.

Petty Spurge *Euphorbia peplus* (7) is a short green annual with oval, blunt-tipped leaves. The flowers (see Wood Spurge above) have horned glands; April onwards. A common weed of cultivation throughout Britain, but becoming rarer in Scotland.

Dog's Mercury *Mercurialis perennis* (8) is a short, unbranched, hairy perennial, with narrow oval leaves in opposite pairs. The male and female flowers occur on separate plants; both are minute and borne in clusters on long stalks; January to May. Frequently forming extensive patches in shady woods, especially beechwoods, but also found in hedges and on mountain rocks. Common except in the extreme N.

MILKWORT FAMILY Polygalaceae
Common Milkwort *Polygala vulgaris* (1). A short perennial with woody stem-bases and upper leaves longer than lower. All the leaves alternate up the stem. The oddly shaped flowers may be blue, purple–pink, white, or white tipped purple; May to September. A common grassland plant throughout Britain. **Heath Milkwort** *P. serpyllifolia* has its lower leaves opposite one another and closer together; it prefers acid soils.

BALSAM FAMILY Balsaminaceae
Himalayan Balsam *Impatiens balsamifera*. A very tall annual with fat, hollow, reddish stems, bearing opposite pairs or whorls of 3 leaves. The large purple–pink flowers, up to 40 mm, have a short spur; July to October. The fruits are explosive. Introduced from the Himalayas and now a common plant by rivers and canals and in wet woods.

Orange Balsam *Impatiens capensis* (2). A short annual with alternate leaves and orange, often red-spotted flowers, 20–30 mm; July to September. The flowers may fail to open. Introduced from N America and common by rivers and canals in S and E England. **Small Balsam** *I. parviflora* (3) has smaller, pale yellow flowers, 5–15 mm; June to October. Naturalised under trees.

MALLOW FAMILY Malvaceae have lobed leaves and large, pink, 5-petalled flowers with many stamens.
Musk Mallow *Malva moschata*. A tall perennial whose leaves are rounded and shallowly lobed at the base and increasingly dissected up the stem. The clear rose–pink or white flowers, 30–60 mm, are borne singly in leafy spikes; July to August. Widespread in grassy places, preferring dry soils, and commonest in the S.

Common Mallow *Malva sylvestris* (3). A tall perennial with almost unlobed basal leaves and usually 5-lobed, ivy-shaped stem-leaves. Most leaves have a dark spot where the leaf joins the stalk. Flowers purplish pink with darker stripes, 25–40 mm; June to October. Common in waste places and roadsides, particularly in the S and E. **Dwarf Mallow** *M. neglecta* is a usually creeping annual with smaller, paler flowers.

Tree Mallow *Lavatera arborea*. A very tall, woody biennial with 5–7-lobed leaves, often folded along the midribs of the lobes. Flowers purplish pink with darker marks, 30–40 mm; June to September. On rocks by the sea in SW England, Wales and the Isle of Man, and introduced elsewhere.

ST JOHN'S WORT FAMILY Guttiferae have opposite leaves and 5-petalled yellow flowers with many stamens.

Tutsan *Hypericum androsaemum*. A semi-evergreen shrub, rarely exceeding 1 m. The broad oval, unstalked leaves are faintly aromatic. Flowers have sepals, petals, and stamens all about the same length and 15–25 mm in diameter; June to August. Fruit a fleshy berry, ripening from red to purple–black. Scattered, but often abundant in woods and hedges, usually within 25 miles of S or W coasts.

Perforate St John's Wort *Hypericum perforatum*. A tall perennial whose upright stems have 2 raised lines. The narrow, unstalked leaves have many small, translucent dots. Flowers 20 mm, petals edged with black dots, sepals narrow, pointed and much shorter than petals; June to September. Common in grasslands and open woods, especially on calcareous soils, except in the mountains and far N.

Slender St John's Wort *Hypericum pulchrum*. A short or tall perennial with round stems. The oval leaves have translucent dots and rolled edges, and clasp the stems. Petals reddish in bud, edged with black dots; flowers 15 mm, June to September. Common throughout Britain in heathland and open woods on acid soils.

Marsh St John's Wort *Hypericum elodes*. A low, creeping, grey-hairy perennial, with rounded, unstalked leaves. Flowers 15 mm but often only half-open, sepals reddish; July to August. A decreasing plant of marshes and wet stream-sides, now largely confined to S and W England and Wales.

Trailing St John's Wort *Hypericum humifusum*. A creeping perennial with very narrow stems. The leaves have translucent dots. The 10 mm flowers are smaller than most other St John's worts, the petals scarcely longer than the sepals. A widespread plant of heaths and moors on acid soils.

DAPHNE FAMILY Thymelaeaceae
Spurge Laurel *Daphne laureola* is an evergreen shrub, often with many little-branched stems. The leaves are narrow, dark glossy green and leathery, and clustered at the top of the stems. Flowers are yellow–green in small clusters; January to April. Fruit a black berry. Widespread in woods on calcareous soils in England, rare in Wales and occasionally introduced in Scotland.

VIOLET FAMILY Violaceae have 5 unequal petals, 2 above and 1 below, the middle of these with a spur.

Sweet Violet *Viola odorata* (1) is a low perennial with long runners, which produces small, rounded leaves in spring and larger, heart-shaped leaves in summer. Flowers 15 mm, rich violet or white, fragrant and with blunt sepals; February to May and sometimes again in autumn. Common in woods and hedges, rarer in the N and W. **Hairy Violet** *V. hirta* has no runners, narrower leaves with hairy stalks, and paler, scentless flowers. Chalk and limestone grassland.

Dog Violet *Viola riviniana* (2) is a low perennial, very variable in size, leafiness and colour. The leaves are heart-shaped and the flowers usually blue–violet, with a paler spur and pointed sepals; March to May and sporadically thereafter. The commonest violet in woods, hedges and grassy places. **Heath Dog Violet** *V. canina* has bluer flowers and narrower, tapering leaves.

Marsh Violet *Viola palustris* is a low, creeping perennial with almost circular leaves, most prominent after flowering. The flowers are small, 10–15 mm, pale lilac with darker streaks, and have a short spur and blunt sepals; April to June (July in mountains). Common in bogs, fens, marshes, and wet woods throughout Britain, except in E Anglia and the Midlands.

Wild Pansy *Viola tricolor* (3) is very variable. Inland usually a short annual with leaf-like stipules where leaves join stem, and mainly blue–violet flowers, often with lowest petal yellow, 15–25 mm; April to November. This is subspecies *tricolor*, found throughout Britain on disturbed sandy soils. Subspecies *curtisii* grows on coastal sand-dunes and is a low perennial with often all-yellow flowers, 10–20 mm.

Field Pansy *Viola arvensis* is a very variable low annual, sometimes hard to distinguish from Wild Pansy. The stipules have broader teeth and the flowers are smaller, 10–15 mm, and usually cream-coloured, the lowest petal with an orange–yellow mark. Petals usually shorter than sepals; April onwards. A common cornfield weed throughout Britain.

ROCK-ROSE FAMILY Cistaceae
Rock-rose *Helianthemum nummularium* is a short undershrub with oblong leaves, green above and grey or white underneath. Flowers 20–25 mm with crinkly petals; June to September. A characteristic plant of chalk and limestone grassland and basic rocks, except in the extreme W.

61

WILLOWHERB FAMILY Onagraceae have 4 petals, 8 stamens, and a long seed-pod containing many cottony seeds.

New Zealand Willowherb *Epilobium nerterioides* is a low, creeping perennial which roots all along the stem. The small, rounded leaves have short stalks. The flowers are tiny, 3–4 mm, on long stalks rising up from the creeping stems, and sometimes not opening; June to August. Introduced from New Zealand and now the commonest dwarf willowherb in upland areas, by streams and on wet rocks; occasional elsewhere.

Rosebay Willowherb *Epilobium angustifolium* is a tall perennial with stiffly erect, leafy stems. The leaves are large and narrow and spirally arranged, and the large rose–purple flowers, 20–30 mm, have unequal petals. The long seed-pods break to release the cottony seeds. Very common and forming extensive patches in woods and heaths and on waste ground.

Great Willowherb *Epilobium hirsutum* (1) is a tall or very tall perennial with conspicuously hairy stems and large, unstalked hairy leaves. The large, purplish pink flowers are up to 25 mm across and have 4-lobed stigmas; July to September. Seed-pods are very long, up to 80 mm. Common, often dominant, by ditches and in damp places, throughout Britain. **Hoary Willowherb** *E. parviflorum* is smaller, with paler flowers.

Marsh Willowherb *Epilobium palustre* is a delicate, short perennial with round stems and narrow leaves. The pale pink flowers, 4–6 mm, often nod in bud and have round-topped stigmas; July to August. Seed-pods 50–80 mm long. In wet places on acid soils and peats throughout Britain, but infrequent in the SE.

Broad-leaved Willowherb *Epilobium montanum* (2) is a short or tall perennial with a rounded, hairless stem and short-stalked, rather broad oval leaves. The flowers are deep pink and drooping in bud, pale when open, 6–9 mm across, and with 4-lobed stigmas; June to September. The commonest willowherb in woods and hedges, and as a weed throughout Britain. **Square-stemmed Willowherb** *E. tetragonum* has square stems and rounded stigmas; in damper places.

LOOSESTRIFE FAMILY Lythraceae
Enchanter's Nightshade *Circaea lutetiana*, a short or tall perennial, is sometimes softly hairy, sometimes hairless, with pointed leaves in opposite pairs. The flowers, 4–8 mm, are in long, leafless spikes; each flower has 2 deeply-lobed petals; June to August. The bristly fruits are on down-pointing stalks. Common in woods on deep soils.

Large-flowered Evening Primrose *Oenothera erythrosepala* is a tall red-hairy biennial with large, spear-shaped, wrinkled leaves, often with a red midrib. The large, yellow, cup-shaped flowers, 80–100 mm across, are pointed in bud and have red-striped sepals; June to September. Introduced and naturalised in dry places, such as sand-dunes and railways.

Purple Loosestrife *Lythrum salicaria* is a tall, softly hairy perennial with narrow, unstalked leaves, mostly in opposite pairs or threes. The flowers are in leafy whorls in a long spike, 10–15 mm across, 6-petalled and bright rose–purple; June to August. Conspicuous among tall vegetation by lakes and rivers, but rare in the N.

CARROT FAMILY Umbelliferae have small 5-petalled flowers in umbels – heads in which all stalks are about the same length and support 1 flower or group of flowers.
Marsh Pennywort *Hydrocotyle vulgaris* is a low, creeping perennial which roots at the leaf-junctions. The leaves are almost entirely round, dimpled in the centre, and with the stalk joining from below. The flowers are tiny, 1–2 mm, pinkish green, and often hidden under the leaves; June to August. Common in damp places and in still, shallow water throughout Britain.

Sanicle *Sanicula europaea* is a perennial with several deeply 5-lobed leaves arising from the base of the often reddish stem. The pinkish or greenish flowers are in clusters on long stalks, in groups of 3, and usually with 3–5 such groups forming a loose flower-head; May to August. Forms extensive patches in dry woods, often on base-rich soils, but becoming rare in the N.

Sea Holly *Eryngium maritimum* is an unmistakable, short, spiny, blue perennial. The lower leaves are rounded and 3-lobed, the upper 5-lobed and unstalked, all with spiny teeth. The bluish flowers are in rounded heads, 15–25 cm across; June to August. On sand-dunes and dry shore-lines except in the NE.

Alexanders *Smyrnium olusatrum* is a tall biennial whose leaves have 3 branches, each with three 3-lobed, dark green leaflets. The yellow–green flowers are in rather tight umbels; April to June. Introduced, but now widely established in hedges and waste places, especially near the sea.

Cow Parsley *Anthriscus sylvestris* (1) is a tall, softly hairy biennial or perennial with ridged stems and large twice or thrice pinnate leaves. Umbels with upper but no lower bracts, bearing white flowers; April to June and sporadically thereafter. Fruit 5 mm, black, with spreading styles. The commonest white umbellifer in spring in hedges and on roadsides. **Rough Chervil** *Chaerophyllum temulentum* has purple-spotted stems and flowers later.

Hedge Parsley *Torilis japonica* is a short or tall annual with delicately twice or thrice pinnate leaves with rounded segments. The white or pink flowers have both upper and lower bracts and outer petals longer than the inner ones; July to September. Common on roadsides and hedges, becoming rare in the NW.

Pignut *Conopodium majus* is a short perennial with an edible swollen root. The root-leaves are short-lived and finely divided, those on the stem more delicate, with linear lobes. Flowers white, 1–2 mm, in lax umbels up to 60 mm across, nodding in bud. Fruit with erect styles. Common on poor soils, typically in dry woods and fields.

Sweet Cicely *Myrrhis odorata* is a tall perennial, emitting a powerful smell of aniseed when crushed. The leaves are large, dark green, and twice or thrice pinnate, their stalks sheathing the stem. Flowers have unequal petals, 2–4 mm, some hermaphrodite in stout umbels, some male only in more slender ones; May to June. Fruits are conspicuously large. Common on roadsides in the N.

Hogweed *Heracleum sphondylium* (2) is a massive, tall or very tall biennial with hairy, hollow stems. The hairy leaves are usually only once pinnate, the large leaflets more or less deeply toothed. The flowers, up to 10 mm across, are sometimes pinkish, the outer ones with very unequal petals; May to October. Very common in many habitats. **Giant Hogweed** *H. mantegazzianum* is huge (often over 4 m) and **dangerous**, contact causing the skin to blister in sunlight.

Hemlock *Conium maculatum* is a very tall biennial with characteristically purple-spotted stems. The leaves are very large but delicate, up to 4 times pinnate, with diamond-shaped leaflets. The flowers are small, 1–2 mm, in large, dense umbels, with upper bracts only on one side; June to August. Common in waste places and by water, becoming rare in the N. **Very poisonous.**·

Angelica *Angelica sylvestris* is a tall to very tall perennial with hollow, often purplish stems. The lower leaves are large and twice or thrice pinnate, the upper extremely reduced and on greatly inflated purplish stalks, surrounding the flower-stalks. Flowers, 2 mm, in umbels up to 150 mm across; July to September. The fruits have broad wings. Common in damp woods and meadows everywhere.

Greater Burnet Saxifrage *Pimpinella major* is a tall perennial with a strongly ridged stem, whose simply pinnate leaves have stalked leaflets. The small white flowers, 2–3 mm, have conspicuous styles and are borne in flat-topped umbels, 30–60 mm across; June to August. It is found in grassy places and open woods, mainly in E England.

Burnet Saxifrage *Pimpinella saxifraga* is a tall, elegant perennial, with a rounded, scarcely ridged stem. The lowest leaves are simply pinnate with rounded leaflets, the next twice pinnate, and the topmost simply pinnate with narrow leaflets. The flowers have very short styles and no bracts; June to September. Widespread in dry grassland except in NW Scotland, but not on acid soils.

Fennel *Foeniculum vulgare* is a tall, greyish perennial with much-divided, feathery leaves with thread-like lobes. The flowers are yellow and have no bracts; July to October. Hedges and disturbed ground, usually near the sea, mainly in the S.

Ground Elder *Aegopodium podagraria* is a short or tall perennial with long, white, underground runners, so that it forms extensive patches. It has large trefoil leaves, the leaflets themselves being trefoil and toothed, forming a level canopy. Flowers are minute, in bractless umbels; June to August. Probably introduced, but very common in disturbed ground.

Wild Parsnip *Pastinaca sativa* is a tall, softly hairy, strong-smelling biennial, and has simply pinnate leaves with lobed leaflets. Flowers are yellow, in large bractless umbels; June to September. The fruits are winged and egg-shaped. The commonest yellow umbellifer in S and E England, mainly on chalk and limestone; elsewhere a relic of cultivation.

poisonous

Rock Samphire *Crithmum maritimum* is a short, rather greyish perennial with fleshy, once or twice trefoil leaves. The flowers are pale greenish yellow and the umbels have conspicuous upper and lower bracts; June to September. The fruit is egg-shaped, turning purplish when ripe. On sea cliffs and rocks, rarely on shingle, along S and W coasts.

Hemlock Water Dropwort *Oenanthe crocata* is a tall or very tall perennial with hollow, grooved stems, and large 2–4 times pinnate leaves. The leaf-stalks form sheaths around the stem. Flower-umbels have both upper and lower bracts, though these soon fall; June to August. The fruits retain conspicuous styles. In damp places and ditches, mainly in the west. **Very poisonous.**

Fool's Parsley *Aethusa cynapium* is a short or tall, hairless annual. The leaves are 2–3 times pinnate, but with few branches. Flower umbels have long, conspicuous upper bracts; June to September. The fruits are egg-shaped and ridged. A common weed in the S, rare in the N.

Pepper Saxifrage *Silaum silaus* is a short to tall, rather delicate perennial. The leaves are 2–3 times pinnate with linear, finely toothed lobes, mostly at the stem-base. The flowers are dull yellow with upper bracts but usually no lower ones; July to September. Fruit egg-shaped. In grassland, rare in the N and W.

Wild Carrot *Daucus carota* is a tall, roughly hairy biennial, with 3 times pinnate leaves. The umbels have conspicuous lobed bracts and are flat-topped, usually becoming concave in fruit, often with a red centre; June to September. Fruits are bristly. In grassy places, particularly near the sea and on chalk, and as a relic of cultivation.

Fool's Watercress *Apium nodiflorum* (1) is a rather floppy perennial whose simply pinnate leaves have shiny, bright green, toothed leaflets. Individual flowers are minute in umbels with upper but no lower bracts; June to September. Common in slow or still water, rare in Scotland. **Lesser Waterparsnip** *Berula erecta* has finely divided underwater leaves and conspicuous lobed bracts.

66

PRIMROSE FAMILY Primulaceae have 5-petalled flowers (except for Chickweed Wintergreen, p.68), and simple leaves.

Primrose *Primula vulgaris* is a low perennial whose large, oval leaves are hairy only on the underside and taper gradually on to the stalk. The flowers, 20–30 mm, are borne singly on long, hairy stalks; February to May. Pink-flowered forms occur as garden escapes. Widespread and often abundant in woods and hedgebanks and, in the W, on sea cliffs.

Cowslip *Primula veris* is a low perennial with hairy leaves, contracting abruptly where they join the stalk. The deep yellow, orange-spotted, bell-shaped flowers, 10–15 mm, are in a nodding cluster on a single stalk; April to May. Grows primarily in old grassland, avoiding acid soils. Widespread, but becoming rare in Scotland.

Birdseye Primrose *Primula farinosa* is a low perennial with mealy white stems and undersides of the leaves, which are in a flat rosette. The exquisite lilac-pink flowers, 8–10 mm, are clustered on a common stalk; June to July. Only in damp places in the limestone hills of N England, but often abundant there.

Yellow Loosestrife *Lysimachia vulgaris* is a tall or very tall, softly hairy perennial. Its long, narrow leaves are in whorls of 2–4 and often bear black dots. The yellow flowers, 15–20 mm, are in loose, leafy clusters; July to August. In damp woods and by rivers and lakes, becoming rare in Scotland.

Creeping Jenny *Lysimachia nummularia* is a low, creeping perennial with rounded, hairless leaves in opposite pairs. The flowers, 15–25 mm, resemble inverted bells with pointed lobes, and have long, thick stalks; June to August. It rarely sets fruit in Britain. In hedges and damp grassy places, becoming rare and finally absent in the N.

Yellow Pimpernel *Lysimachia nemorum* is similar to Creeping Jenny (above) but more delicate in appearance, with pointed-tipped leaves and smaller, flat flowers with rounded petals; May to August. The flower-stalks are longer than the leaves and very slender. Occurs throughout Britain in woods and shady or moist hedges and fields.

67

Sea Milkwort *Glaux maritima* is a creeping, succulent perennial with small, oblong, very short-stalked leaves in opposite pairs. The flowers have no petals, only 5 pink sepals, 5 mm across; May to September. A common salt-marsh plant throughout Britain, rarely on rocks and cliffs by the sea.

Water Violet *Hottonia palustris* is a delicate, pale green, aquatic perennial whose pinnate leaves have many linear lobes, giving a feathery appearance. The pale lilac flowers, 20–25 mm, are in whorls on a long, elongating stalk; May to June. A decreasing plant of still fresh water, mainly in E England.

Scarlet Pimpernel *Anagallis arvensis* is a sprawling annual with square stems and oval, unstalked leaves. The pale scarlet, or rarely lilac or blue, flowers, 10–15 mm, are borne singly on long, slender stalks; May to September. The petals are fringed. A widespread arable weed, becoming rare in the N.

Bog Pimpernel *Anagallis tenella* is a delicate, creeping, mat-forming perennial with tiny, rounded, short-stalked leaves in opposite pairs. The elegant, pink, bell-like flowers, 10–15 mm, are borne on long, slender, erect stalks and open in sunshine; June to August. Widespread in damp grassland and dune slacks and on peat, but commonest in the W.

Brookweed *Samolus valerandi* is a short, hairless perennial with unbranched, leafy stems and oval or spoon-shaped leaves. The white flowers, 2–4 mm, have joined petals; June to August. The flower-stalks bear tiny bracts and become bent in fruit. In damp places, especially near the sea, and commonest in the S.

Chickweed Wintergreen *Trientalis europaea* is a short, slender, upright perennial with a single whorl of leaves at the top of the unbranched stem, and a few small leaves below. One or two flowers, 15–18 mm, each with 7 petals, arise from the whorl; June to August. In pinewoods and on moors, common in E Scotland, very rare in E England.

HEATH FAMILY Ericaceae are small shrubs with simple, often leathery leaves and bell-shaped flowers with 4–5 joined lobes.

Heather or **Ling** *Calluna vulgaris* is a short, straggly, evergreen shrub. The main shoots have few leaves and bear short shoots with 4 dense ranks of tiny, linear leaves. Flowers are solitary, bell-shaped, 4–5 mm long, and pale pinkish purple; July to October. Common everywhere on acid soils on heaths, moors, and in open woods, often the only plant for miles.

Bell Heather *Erica cinerea* is a short, hairless, evergreen shrub with leaves in whorls of 3 up the stem, each whorl giving rise to a short, leafy shoot. Flowers are bell-shaped, 5–6 mm long, and deep red–purple, in short, leafy shoots; June to September. On dry heaths and moors, commonest in the N and W.

Cross-leaved Heath *Erica tetralix* is a short, sprawling shrub with grey-downy twigs and leaves. The leaves are small and narrow, in whorls of 4. Flowers are pink, flask-shaped, 6–7 mm long, and in tight clusters at the tips of the twigs; June to October. On wet heaths and moors, common wherever its habitat is found.

Bilberry *Vaccinium myrtillus* is a short, deciduous shrub with green twigs and bright green, oval leaves. Flowers are flask-shaped, greenish pink, 4–6 mm long, and half hidden by the leaves; April to June. Fruits are purplish black in August and September, edible and good. Heaths, moors, and open woods on acid soils, scarce in the SE.

Cowberry *Vaccinium vitis-idaea* (1) is a low or short evergreen shrub with small, rounded, often notched, dark, glossy green leaves. The bell-shaped, pink-tinged white flowers, 5–6 mm long, are in short, leafy spikes; May to July. Fruit is an edible red berry. Moors and woods on acid soils, mainly in the hills. **Cranberry** *V. oxycoccos* (2) is low and creeping with smaller leaves and reflexed petals. Bogs.

CROWBERRY FAMILY Empetraceae **Crowberry** *Empetrum nigrum* is a low or short, heather-like shrub whose red or green stems are crowded with many narrow leaves. The tiny 6-petalled flowers, 1–2 mm, are pink–purple and solitary at the base of the leaves; April to June. Fruit is a black berry. Moorland and mountainsides, only in the hills.

WINTERGREEN FAMILY Pyrolaceae
Common Wintergreen *Pyrola minor* is a low perennial with a rosette of oval, finely toothed leaves. The almost spherical, white or very pale pink flowers, 6 mm, are in a single, unbranched, rather dense spike; June to August. Widespread in the N but very local in the S, in woods, on moors, on rock-ledges on mountains, and in dune-slacks.

SEA-LAVENDER FAMILY
Plumbaginaceae
Thrift *Armeria maritima* is a low perennial forming more or less tight cushions. The very narrow, fleshy leaves can be as much as 150 mm long, and are all in a basal rosette. The flowering stalk bears a cylindrical brown sheath just below the dense globular head of pale to dark pink, or white flowers; April to July and sporadically thereafter. A plant of sea-cliffs and salt-marshes, and more rarely in upland areas inland.

Common Sea-lavender *Limonium vulgare* (1) is a short perennial with a rosette of narrow oval leaves which have prominent pinnate veins. The flowers are in tight clusters within a flat-topped flower-head, which is much-branched from above the middle of its stalk; July to September. In salt-marshes from S Scotland southwards. **Lax-flowered Sea-lavender** *L. humile* is less common. It has loose flower-heads branching from below the middle.

BOGBEAN FAMILY Menyanthaceae
Bogbean *Menyanthes trifoliata* is a short aquatic perennial with large trefoil leaves emerging from the water. The flowers are pink outside and paler within, 15–20 mm, and fringed with long white hairs; April to June. A widespread and locally abundant plant of shallow, nutrient-poor water, but also found in marshes and bogs.

Fringed Water-lily *Nymphoides peltata* is more closely related to gentians than to the true water-lilies (p.40). It is an aquatic perennial with floating rounded leaves, often with purple spots. The flowers have fringed petal-lobes, 30–35 mm; June to September. Found in still and slow fresh water, mainly in E and central England; more often introduced than native.

PERIWINKLE FAMILY Apocynaceae
Lesser Periwinkle *Vinca minor* is a low, scrambling shrub with trailing, rooting stems. The narrow, oval leaves are evergreen and leathery, and the blue–violet flowers, 25–30 mm, are scattered on short, upright stems; February to May. Scattered in woods and hedges throughout Britain, but probably often an escape.

PHLOX FAMILY Polemoniaceae
Jacob's Ladder *Polemonium caeruleum* is a tall, leafy perennial with pinnate leaves, each with up to 12 pairs of rather narrow, pointed leaflets. It has purplish blue or occasionally white flowers, 20–30 mm, in a many-flowered spike; June to August. In open, rocky woods in the hills of N England, and more common as a garden escape.

GENTIAN FAMILY Gentianaceae have flowers with 5 petals joined at the base but the lobes free; leaves opposite.
Common Centaury *Centaurium erythraea* (1) is a low or short annual, very variable in size and vigour and the size and shape of the flower-head. It has a rosette of strongly veined leaves and pink flowers in dense clusters; June to September. A common grassland plant in England and Wales, but almost only on sand-dunes in Scotland. **Lesser Centaury** *C. pulchellum* has no rosette leaves and longer-stalked flowers. Less common in damp places, mainly by the sea.

Yellow-wort *Blackstonia perfoliata* is a short, rather greyish annual with a rosette of leaves and opposite pairs of more triangular leaves joining around the stem. It has yellow, 6–8 petalled flowers, 10–15 mm, in a loose cluster; June to October. Frequent in chalk and limestone grassland in S and central England, becoming scarce further N.

Marsh Gentian *Gentiana pneumonanthe* (2) is a short, unbranched perennial with small, narrow leaves and small clusters of large, bright blue, trumpet-shaped flowers, 25–40 mm long, and striped green outside; July to September. It is a scarce and decreasing plant of wet heathlands in a few places in England. **Field Gentian** *Gentianella campestris* (3) has bluer, 4-lobed flowers. Commoner in the N.

Autumn Gentian *Gentianella amarella* is a short biennial with a rosette of leaves in its first year, but stem-leaves only on flowering plants. The flowers are dull purple, 14–22 mm long, with a fringed tube and 4 or 5 spreading lobes; July to September. A widespread plant of dry basic grassland, commonest in the S.

Spring Gentian *Gentiana verna* (4) is a low, tufted perennial with a rosette of oval leaves and a leafy flowering stem. Each stem bears a single bright blue flower, 15–25 mm long, with 5 spreading lobes; May to June. A very rare plant of short turf on limestone, in and around upper Teesdale. **Alpine Gentian** *G. nivalis* is a delicate annual with smaller flowers and only grows in the Scottish Highlands.

BINDWEED FAMILY Convolvulaceae are mainly climbers, twisting anti-clockwise. Flowers trumpet-shaped (except Dodder).

Hedge Bindweed *Calystegia sepium* is a very tall, climbing perennial with stems that twine anti-clockwise and large arrow-shaped leaves. The large white flowers, 30–40 mm, are borne singly, and have 2 large bracts surrounding the sepals; June to September. Extremely common in the S, becoming rarer in the N, both in natural vegetation – hedges woods, and fens – and as a weed.

Field Bindweed *Convolvulus arvensis* is a creeping or occasionally climbing perennial with arrow-shaped, sometimes greyish leaves, and stems twining anti-clockwise. The flowers are pink, often with paler stripes, 15–30 mm across, and have very small bracts; June to September. A widespread and persistent weed in lowland areas and in the S, also found frequently by paths.

Sea Bindweed *Calystegia soldanella* is a short perennial with long underground stems and long-stalked, kidney-shaped leaves. The white-striped pink flowers, 40–50 mm across, have conspicuous bracts that are shorter than the sepals; June to August. A plant of coastal sand, typically among Marram Grass on mobile dunes; mainly on S and W coasts, rarer on the E coast.

Common Dodder *Cuscuta epithymum* (1) is an extraordinary climbing, leafless parasite with clusters of pink flowers on the red stems. It obtains all its nutrition from its host, which is usually either Gorse (p.52) or Heather (p.69). Uncommon and most often found on SW heaths. **Greater Dodder** *C. europaea* (2) has bigger heads of larger, blunt-petalled flowers. Rarer and usually on nettles.

BEDSTRAW FAMILY Rubiaceae have square stems, whorls of narrow leaves, and clusters of 4-petalled, star-shaped flowers.

Field Madder *Sherardia arvensis* is a short, weak-stemmed, hairy annual with whorls of from 4 (at the base of the stem) to 6 (at the top of the stem) narrow oval leaves. The pale pinkish-purple flowers, 3 mm, are in heads surrounded by narrow, leaf-like bracts; May to September. A common plant of disturbed ground, particularly arable fields, in the S, becoming scarce further N.

Squinancywort *Asperula cynanchica* has several delicate, short, 4-angled stems, rising from a perennial stock. The leaves are very narrow, usually 4 in a whorl, but varying in length. The flowers are small, 3–4 mm across, and pale pink, though white inside the tube. They are borne in lax clusters; June to September. Dry grassland, mainly on chalk in the S of England.

Wild Madder *Rubia peregrina* is a tall, scrambling evergreen with rough, square stems whose downward-pointing prickles help the plant to clamber over other vegetation. Its shiny, dark green leaves are in whorls of 4–6 and are leathery and prickly. The small yellow–green flowers, 5 mm, have 5 lobes and are borne in numerous clusters; June to August. The fruit is a black berry. In scrubby vegetation in the SW.

Woodruff *Galium odoratum* is a short perennial with unbranched square stems bearing whorls of 6–8 narrow leaves, with tiny prickles along their edges. The flowers have 4 lobes, about 6 mm across, and are in long-stalked clusters; April to June. The fruits are covered with prickly bristles. A widespread woodland plant, often forming large patches on deeper soils.

Lady's Bedstraw *Galium verum* is a short, much-branched, rather sprawling perennial, whose stems turn black when dry. The very narrow leaves have their margins rolled in underneath and are in whorls of 8–12. The flowers are bright yellow, 2–4 mm, copiously produced in leafy clusters; June to September. A common plant of dry grassy places, except on very acid soils.

Hedge Bedstraw *Galium mollugo* (1) is a tall, scrambling perennial with smooth, square stems, and oval leaves in whorls of 6–8, each with 1 vein and a pointed tip. The white flowers, 2–3 mm, have pointed petals; June to September. The fruits are dark green becoming black. A common plant of hedges and scrubland in S England, becoming scarce in the N and rare in Scotland. **Heath Bedstraw** *G. saxatile* is a less robust plant of acid grasslands and heaths; it has triangular petals.

Common Cleavers *Galium aparine* is a tall, scrambling annual entirely covered with tiny, downward-pointing prickles, so that all parts cling to fur or clothing. The leaves are in whorls of 6–8 and the inconspicuous whitish flowers, 2 mm, have long, straight stalks projecting from the whorls; May to September. A very common plant growing in a wide range of disturbed and undisturbed habitats.

Crosswort *Cruciata laevipes* is a short, hairy perennial with yellowish, 3-veined, oval leaves in whorls of 4. The pale yellow flowers, 2 mm, are in short clusters, half hidden by the leaves, and some are only male; April to June. The fruit is greenish black. Hedges and grassland, often on limestone; commonest in N England, less so in the S, and rare in N Scotland.

1

BORAGE FAMILY Boraginaceae are mostly roughly hairy with simple leaves and coiled, 1-sided spikes of 5-petalled tubular flowers, typically pink in bud and blue when open.

Bugloss *Lycopsis arvensis* is a short annual covered with rough hairs. It has rather narrow leaves with crinkly edges, and bright blue flowers, 5–6 mm across, in long clusters, that sometimes branch; May to September. Bugloss is an arable weed also found in sandy places near the sea, primarily in the E but scattered elsewhere.

Oysterplant *Mertensia maritima* is a creeping perennial forming spreading, blue–grey mats. It has 2 rows of fleshy oval leaves and loose clusters of flowers, 6 mm, that open pink and turn blue–purple; June to August. It grows on coastal shingle in N Scotland and occasionally as far S as Lancashire.

Houndstongue *Cynoglossum officinale* is a tall, softly hairy, greyish biennial that smells of mice. It has long, rather narrow leaves, silky with flattened hairs, and clusters of maroon flowers, 8–10 mm; June to August. It grows in dry, grassy places in S England and on sand dunes near the sea further N.

Common Comfrey *Symphytum officinale* (1), a tall, softly hairy perennial, has large, oval leaves on winged stalks. The flowers are either pink–purple or cream and tube-like, 15–18 mm long, with a conspicuous style; May to June. They are borne in curled, nodding clusters. A widespread and common plant, usually found near fresh water.
Russian Comfrey *S.* × *uplandicum* (2) has stiff hairs and bluish flowers and is common in waste places.

Common Gromwell *Lithospermum officinale* (3) is a tall, rough-stemmed perennial bearing clusters of flowers on several branches. It has long, narrow leaves with conspicuous lateral veins. The flowers are creamy white and 3–4 mm across; May to July. Common Gromwell is found in open woods and bushy places, usually on limestone soils, in S and E England.

Corn Gromwell *Buglossoides arvensis* is a short annual with rough stems and narrow, 1-veined leaves. The stems are usually unbranched except at the top. It bears tight clusters of creamy white flowers, each 3–4 mm, little longer than the sepals and much shorter than the conspicuous bracts; May to August. It is a frequent arable weed in SE and central England, but scarce elsewhere.

cream variety

1

3

2

Field Forgetmenot *Myosotis arvensis* is a short, softly hairy annual with a rosette of stalked leaves, which sometimes over-winters, and leafy flowering stems bearing curved spikes of small blue flowers, up to 5 mm across; April to September. The flower-stalks lengthen in fruit. A common arable weed, also found in open woods and on sand dunes, throughout Britain.

Water Forgetmenot *Myosotis scorpioides* is a short, rather floppy perennial, with a creeping stem producing more or less up-right flowering shoots. It looks hairless, but on close examination has short, flattened hairs. The leaves are oblong and tapering to the base, and the sky-blue flowers (rarely pink or white) are up to 10 mm across; June to September. Common everywhere in wet places.

Wood Forgetmenot *Myosotis sylvatica* is a short perennial with conspicuous spreading hairs and a rosette of oval leaves. The leaves on the stem are narrower, and the bright blue flowers, 6–10 mm across, are in loose spikes that elongate greatly in fruit; April to July. The sepal-tube is covered with hooked hairs. A local plant, mainly in up-land woods in N England; rare in Scotland.

Changing Forgetmenot *Myosotis discolor* (1) is a delicate, often diminutive annual, covered in hairs and with narrow oblong leaves. The flowers are in short, compact spikes and open off-white, turning to blue; May to July. Each flower is only 2–3 mm across. Widespread but not common on sandy soils. **Early Forgetmenot** *M. ramosissima* has all-blue flowers, and longer flower-spikes.

Green Alkanet *Pentaglottis sempervirens* is a tall, rather coarse perennial that feels rough to the touch. It has large, deep green, pointed oval leaves, and bright blue flowers, 10 mm across, in dense, long-stalked, leafy spikes; April to July. It is most probably an introduced plant and is found here and there throughout Britain, mainly in hedges; commonest in the SW.

Viper's Bugloss *Echium vulgare* is a very rough, hairy perennial ranging from short to tall. It has long, rather narrow, wavy-edged leaves with apparently no side-veins. The bright blue, tubular flowers, 15–20 mm long, are in dense spikes, clustered at the tip of the stem; May to September. The style and 4 stamens project from the tube. In sandy grassland; common in the S, be-coming rare further N.

VERBENA FAMILY Verbenaceae
Vervain *Verbena officinalis* is a short, rather inconspicuous perennial with thin, square, tough stems. The lower leaves are deeply toothed but the upper ones are narrow and untoothed. The dull lilac-coloured flowers, 3–4 mm across, appear almost 2-lipped; June to October. The flower-spikes are crowded at first but elongate in fruit. Common by roadsides in S England but rarely found N of the Humber.

LABIATE FAMILY Labiatae have square stems and opposite leaves. Flowers 2-lipped (except Bugle and Wood Sage). Often aromatic.
Bugle *Ajuga reptans* has long creeping runners producing upright, short, leafy flowering stems. The lower leaves are large and long-stalked and form a rosette, while the upper ones are unstalked. Bugle has distinctive pale blue flowers with a thin upper lip and large 3-lobed lower lip; April to June. A common plant of damp woods and meadows.

Skullcap *Scutellaria galericulata* is a short, square-stemmed perennial with pairs of round-toothed, oval leaves, rather squared at the base. The flowers are blue–violet and 2-lipped, the tube up to 20 mm long, but are scattered in pairs up the stem; June to September. A widespread plant of wet, grassy places and riversides.

Wood Sage *Teucrium scorodonia* is a short perennial with wrinkled, dull yellow–green, heart-shaped leaves. It produces long, al-most leafless spikes of soft greenish yellow flowers, each 8–10 mm long, with no upper lip; July to September. It is common throughout Britain in a wide range of habi-tats including woods, rocks and dunes, but always on acid soils.

Self-heal *Prunella vulgaris* is a low or short perennial with leafy stems bearing pairs of oval leaves and each ending in a dense cluster of violet flowers. The flowers are 2-lipped, the upper lip very hooded; June to October. A very common grassland plant, on all but the most acid soils; also found in open woods.

Ground Ivy *Glechoma hederacea* is a softly hairy creeping perennial which smells strongly when crushed. It has rounded, heart-shaped leaves on long stalks, and whorls of 2–4 2-lipped violet flowers, each 15–20 mm long, all up the stem; March to June. A widespread and common plant of woods, hedges, grassland, and disturbed ground, becoming rare in N Scotland.

76

White Dead-nettle *Lamium album* is a creeping perennial with short, upright, hairy stems. It has toothed, heart-shaped leaves, similar to those of Nettle (p.30) but without stinging hairs. The 2-lipped white flowers, 20–25 mm long, are in whorls at the base of the leaves; March to November. A common weed of disturbed ground, becoming scarce in Scotland.

Red Dead-nettle *Lamium purpureum* (1) is a short, usually well-branched and purple-tinged annual. It has long-stalked, toothed, heart-shaped leaves, and whorls of 2-lipped pinkish purple flowers, 10–15 mm long; all year. A very common weed of disturbed ground. **Henbit Dead-nettle** *L. amplexicaule* has unstalked upper leaves. It is a common weed in lowland areas.

Common Hemp-nettle *Galeopsis tetrahit* is a short to tall annual with rough, hairy stems, swollen where the leaves join. The leaves are toothed and stalked, and the flowers are in whorls with a conspicuous 5-lobed pointed sepal-tube which persists after the petal-tube has fallen. The petal-tube is 2-lipped, pinkish purple, 15–20 mm; July to September. A widespread and common plant, both as an arable weed and in woods and wet, peaty places.

Large-flowered Hemp-nettle *Galeopsis speciosa* is very similar to Common Hemp-nettle but is usually larger and has very distinctive flowers: the upper lip is pale yellow and the lower one purple, with orange–yellow markings near the throat. A widespread plant of disturbed ground, most often on peaty soils; almost absent from S England.

Yellow Archangel *Lamiastrum galeobdolon* is a creeping, sometimes mat-forming perennial, with short, upright flowering stems. It has dark green, toothed, oval leaves, broader on the runners than on the flowering stems. The 2-lipped yellow flowers, 20 mm long, are in whorls at the base of the leaves; May to June. A frequent woodland plant in S and central England, becoming rare further N.

Black Horehound *Ballota nigra* is a tall, untidy perennial with a strong and rather unpleasant smell. It has hairy, oval leaves, and whorls of purple–pink flowers, 12–18 mm long; June to September. The sepal-tube has spreading teeth. A common plant of disturbed ground in England, becoming rare in the N and W.

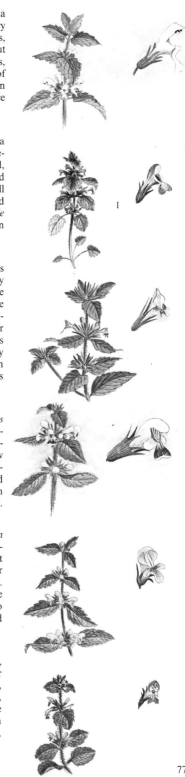

1

Betony *Betonica officinalis* is a short, slightly hairy perennial all of whose oblong, round-toothed leaves have long stalks, except for the top pair. The bright red-purple, 2-lipped flowers, 15 mm long, are in dense, leafy spikes; June to October. A common grassland plant, mainly on chalky and sandy soils, in England and Wales; rare in Scotland.

Hedge Woundwort *Stachys sylvatica* (1) is a tall, roughly hairy perennial with a strong smell when crushed, unpleasant to most people. The toothed, heart-shaped leaves have long stalks, and the deep wine-purple flowers, 12-15 mm long, have white markings on the lower lip; June to October. A very common plant of hedges and open woods. **Marsh Woundwort** *S. palustris* has paler flowers and narrower, unstalked leaves. On damp soils.

Basil Thyme *Acinos arvensis* is a low, weak-stemmed annual with slightly hairy, short-stalked, oval leaves. The violet flowers, 8-10 mm long, have a conspicuous white mark on the lower lip and are arranged in whorls of 3–8, usually 6; June to September. It grows on disturbed ground on dry limestone soils, widely in S England, but is rare in the N and W.

Calamint *Calamintha sylvatica* is a short, mint-scented, hairy perennial, with long-stalked, oval leaves. The dark-spotted, pale lilac flowers, 10-15 mm long, are borne in clusters of up to 10 flowers at each leaf-junction; July to September. Calamint is a local plant of dry grassland on chalk and limestone, N as far as Yorkshire.

Wild Clary *Salvia horminoides* is a tall perennial, with hairy, often reddish stems and large, oval lower leaves, which may be deeply lobed or only shallowly toothed. The flowers are in long, whorled spikes, some large (up to 15 mm long), open, 2-lipped and blue–violet, but others small and never opening; June to September. Dry grassland and roadsides in S and E England.

Meadow Clary *Salvia pratensis* is a tall, hairy perennial with long-stalked, toothed lower leaves and a long, whorled spike of brilliant blue flowers, each up to 25 mm long; June to July. Some plants have smaller flowers as well. Meadow Clary is a rare native of dry limestone grassland, but is also more widely found as an introduction S of the Humber.

Water Mint *Mentha aquatica* (1) is a variable short or tall perennial with reddish, hairy stems and toothed, oval leaves. The pinkish lilac flowers, 8-10 mm long, are in a dense terminal head, and usually also 2–3 long-stalked clusters further down the stem. A common plant of wet places, often actually in water. **Corn Mint** *M. arvensis* (2) has tight, leafy whorls of flowers at the base of the leaves. In damp places and disturbed ground.

Spearmint *Mentha spicata* is the mint most often grown in gardens. It is a tall, strong-smelling perennial with rather narrow, unstalked leaves and pale lilac flowers in tight whorls, the topmost ones condensed into a pointed, cylindrical spike; July to October. Most often found by roads and in waste places as an escape from gardens. Several other species and many hybrids occur.

Gipsywort *Lycopus europaeus* is a tall, slightly hairy perennial with narrow, deeply and regularly toothed leaves, the lower ones sometimes pinnately lobed. The flowers are in whorls at the base of the leaves. They have long, pointed lobes to the sepal-tube and a small, whitish, 2-lipped petal-tube, 3 mm; July to September. Common in S and central England in marshes and by rivers, but local in the W and rare in the N.

Wild Basil *Clinopodium vulgare* is a short, rather weak-stemmed, hairy perennial with finely toothed, oval leaves. The flowers are bright pink–purple, 15–20 mm long, with a conspicuous, white-haired, purplish sepal-tube; July to September. The whorls of flowers have protruding, bristly bracts. Scrubby grassland on dry, usually limestone soils, commonest in the S.

Marjoram *Origanum vulgare* is a tall, branching perennial with scarcely toothed, stalked, oval leaves. It has tight clusters of pale purplish pink flowers, 6–8 mm, surrounded by purple bracts; July to September. Marjoram is common in dry, usually limestone grassland in S England, becoming rarer further N.

Thyme *Thymus praecox* is a creeping, woody perennial with short-stalked, oval, aromatic leaves. It has tight heads of pinkish purple flowers, from June to September, but these are often attacked by an insect and replaced by white woolly galls. It is common in dry grassland and heaths except in E and central England.

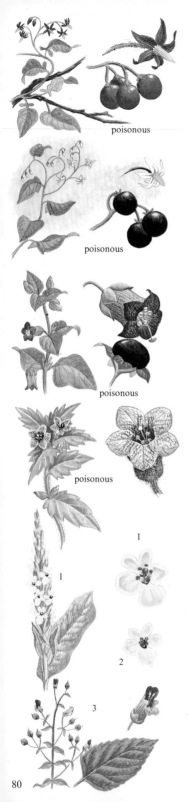

poisonous

poisonous

poisonous

poisonous

1

1

1

2

3

NIGHTSHADE FAMILY Solanaceae have 3 petals, joined at the base, but the lobes free. Many are poisonous.

Bittersweet *Solanum dulcamara* is a woody perennial, sometimes scrambling high up into other vegetation. It has pointed, oval leaves, often with 2 lobes or separate leaflets at the base, and potato-like flowers with bright purple, spreading or turned-back petals, and a conspicuous column of yellow anthers; May to September. The fruit is a **poisonous** oval red berry. Common in damp woods, hedges and fens, on shingle, and as a weed, except in N Scotland.

Black Nightshade *Solanum nigrum* is a short annual with dark stems and pointed, oval, sometimes toothed leaves. It has white flowers, smaller than Bittersweet (5–6 mm), but otherwise similar; July to October. Its **poisonous** berry is initially green but ripens black and spherical. A widespread weed in England and Wales, casual in Scotland.

Deadly Nightshade *Atropa bella-donna* is a tall, thick-stemmed, branching perennial with large, oval leaves. It has dull or greenish purple, bell-shaped flowers, 25–30 mm long, quite unlike those of the 2 nightshades above; June to August. It bears the best-known of **poisonous** berries, glossy black and up to 20 mm across. Open woods on limestone, mainly in the S.

Henbane *Hyoscyamus niger* superficially resembles Deadly Nightshade, but is stickily hairy and foul-smelling. It has oblong leaves and purple-marked flowers with a yellow background; May to September. **Extremely poisonous**; disturbed ground, particularly near the sea, and mainly in the S.

FIGWORT FAMILY Scrophulariaceae have flowers either flat and with 4 (speedwells) or 5 (mulleins) joined petals, or 2-lipped.

Great Mullein *Verbascum thapsus* (1) is a stout, very tall biennial, woolly with thick, white hairs. It has long, rather narrow leaves, in a rosette in the first year and often pressed to the stem in the second. The flat, yellow flowers are 15–30 mm across; June to August. Common on dry soils, except in N Scotland. **Dark Mullein** *V. nigrum* (2) has a ridged stem and dark green leaves; mainly in the S.

Figwort *Scrophularia nodosa* (3) is a tall perennial with a square stem and finely-toothed, pointed, oval leaves. The flowers are 2-lipped, 8–10 mm long, with a brown upper and green lower lip; June to September. Common in damp woods. **Water Figwort** *S. auriculata* has broadly-winged stems. Wetter places and more southern.

Snapdragon *Antirrhinum majus* is a short, sometimes tall perennial whose well-branched stems, woody at the base, form small clumps. The stems bear narrow leaves and end in a spike of bright purplish red or pink flowers, 30–40 mm long, which end in a pouch, not a spur; May to October. A garden plant, naturalised on walls and rocks, mainly in the S; many colour variations occur.

Sharp-leaved Fluellen *Kickxia elatine* (1) has stickily-hairy, trailing stems with long-stalked, pointed triangular leaves. It is an annual and has long-spurred flowers, 7–9 mm, with a pale purple upper lip and a yellow lower lip. The flowers are borne singly at the leaf junctions; July to October. Locally abundant in arable fields in S England. **Round-leaved Fluellen** *K. spuria* has rounded leaves.

Common Toadflax *Linaria vulgaris* has rather bluish, very narrow leaves clothing its often branched stems. It is a perennial with long spikes of 20 or more long-stalked flowers, 15–25 mm, yellow with an orange centre; June to October. Common in hedge-banks and grassy places, but absent from NW Scotland. The seed capsules nearly always contain the caterpillars of a moth.

Pale Toadflax *Linaria repens* is a greyish perennial with a creeping underground stem, producing several upright flowering shoots with narrow leaves. The flowers are pale lilac, with darker veins, 7–14 mm, and have a short spur; June to September. It grows in dry, rather open ground, mainly in the S and W.

Ivy-leaved Toadflax *Cymbalaria muralis* is a trailing perennial with purplish stems and long-stalked, 5-lobed, ivy-shaped leaves. The flowers are borne individually on long stalks from the leaf-junctions, and are lilac with a yellow centre, 8–10 mm; April to November. A garden plant, now extremely widely naturalised on rocks and walls.

Small Toadflax *Chaenorrhinum minus* is a low, upright annual with downy stems and narrow, rounded leaves. It has pale purple flowers with a slightly open mouth between the 2 lips, and a short spur; May to October. The flowers are only 6–8 mm, and are long-stalked, arising from the leaf-junctions. Widespread in dry bare places, particularly railway tracks, north to S Scotland.

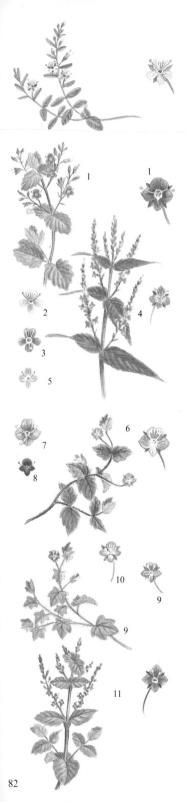

Thyme-leaved Speedwell *Veronica serpyllifolia* is a low, creeping perennial with upright flowering shoots. It has opposite pairs of clear green, rather shiny leaves, and long, loose spikes of very pale blue flowers, 5–6 mm, with darker veins; April to October. Very common in moist grassland, but avoiding peaty soils; more rarely as a weed of bare ground.

Germander Speedwell *Veronica chamaedrys* (1) is a short, floppy-stemmed perennial with 2 lines of hairs down the stems. It has oval, hairy leaves, and bright blue, white-centred flowers 10–12 mm, in long, loose spikes; April to June. Very common in dry grassy places. **Heath Speedwell** *V. officinalis* (2) has paler lilac flowers, dark-veined, 5–6 mm, in dense spikes. **Wood Speedwell** *V. montana* (3) has evenly hairy stems, longer-stalked leaves, and paler flowers, 7 mm. Damp woods.

Water Speedwell *Veronica anagallis-aquatica* (4) is a short, fleshy perennial with narrow leaves clasping the stem. The long, upright spikes of pale blue flowers, 5–6 mm, arise in pairs from the bases of the leaves; June to August. Widespread in mud on the edge of water; scarce in Scotland. **Pink Water Speedwell** *V. catenata* (5) has pink flowers.

Field Speedwell *Veronica persica* (6) is a low, hairy annual, with sprawling stems bearing pairs of pale green, toothed oval leaves. Flowers are borne singly on long stalks arising from the leaf-junctions, each 8–12 mm, the lowest lobe pale. A very common weed of disturbed ground. **Slender Speedwell** *V. filiformis* (7) is creeping and patch-forming, and has very long-stalked flowers. Increasing in lawns and damp grassland. **Grey Field Speedwell** *V. polita* (8) is greyish and has smaller flowers.

Ivy-leaved Speedwell *Veronica hederifolia* (9) is a low, sprawling, hairy annual with pale green, short-stalked, ivy-shaped leaves. It has blue flowers, 6–9 mm, on short stalks arising from the leaf-junctions; March to August. A very common weed of cultivated ground in S England, becoming scarce in Scotland. Subspecies *lucorum* (10), with pale lilac flowers, 4–6 mm, and narrower leaf-lobes, also grows in woods and hedges.

Brooklime *Veronica beccabunga* (11) is a short, fleshy perennial with thick, oval, round-toothed leaves on short stalks. The bright blue flowers, 7–8 mm, are in opposite pairs of rather lax spikes; May to September. The commonest speedwell in and near water, and in damp meadows.

Monkey Flower *Mimulus guttatus* is a short, rather weak-stemmed perennial with rounded, oval toothed leaves. It has large, long-stalked, bright yellow flowers, 25–45 mm, with small red spots in the throat; June to September. Introduced from N America, but now widespread in ditches and on river shingle; not in E England.

Foxglove *Digitalis purpurea* is a tall, downy biennial which forms a large rosette of wrinkled, narrow oval leaves in the first year, and a long flowering stem in the second. The spikes have up to 80 pink–purple, tubular flowers, 40–50 mm; June to September. White-flowered plants are locally frequent. Common in woods and moors and on sea-cliffs, often carpeting the ground the year after trees are felled.

Red Bartsia *Odontites verna* (1) is a short, purplish, much-branched annual, with opposite pairs of narrow, toothed leaves. It has long spikes of pinkish purple flowers, 8–10 mm, with the lower lip 3-toothed; June to September. Parasitic on the roots of other plants in bare and grassy places. **Yellow Bartsia** *Parentucellis viscosa* (2) has yellow flowers, 20 mm. SW England.

Common Cow-wheat *Melampyrum pratense* is a very variable, low or short annual with opposite pairs of long, narrow leaves. The flowers are in pairs at the base of the leaves, both facing the same way; they are yellow, sometimes very pale or purple-marked, 10–15 mm; May to September. Common in open woods, heaths, and poor grassland, except in E England.

Eyebright *Euphrasia officinalis* is an extremely variable collection of very similar species, all low or short annuals. Stems and toothed, oval leaves may be purplish or bronzed, and plants are often well-branched. The flowers are white, variously tinged and veined with purple, 3–15 mm, with the lower lip split into 3 toothed lobes; June to October. Parasitic on the roots of other plants and very common in short grassland.

Yellow Rattle *Rhinanthus minor* is a rather variable, short annual, with a square, black-spotted stem. It has narrow, oblong, evenly-toothed leaves and yellow flowers, 10–20 mm, in loose or dense, leafy spikes; May to September. The sepal-tube forms an inflated sack round the seeds, which rattle inside when ripe, hence the name. Parasitic on the roots of other plants in undisturbed grassland, and occasionally found in cornfields.

Lousewort *Pedicularis sylvatica* is a low, bushy perennial, branching frequently from the bottom of the stem. It has pinnate leaves with deeply toothed leaflets, giving a feathery appearance, and bright pink flowers, 20–25 mm, with a pair of teeth on the upper lip; April to July. Common in damp heaths and moors, except in E England.

Marsh Lousewort *Pedicularis palustris* is a low to short annual with a single stem, often branching from near the middle. It has reddish pink flowers, 20–25 mm, with 2 pairs of teeth on the upper lip, and a downy sepal-tube; May to September. Wet heaths and damp grassland; common in the N, very scarce in the SE.

Cornish Moneywort *Sibthorpia europaea* (1) is a low, creeping perennial with rounded leaves and tiny, solitary, pink and yellow flowers; June to October. It is a rare plant of damp places in the SW. **Fairy Foxglove** *Erinus alpinus* (2) is a low, hairy perennial forming neat tufts. It has spoon-shaped, toothed leaves, in a tight rosette, and scattered up the flowering stems, and purple flowers with 5 notched petal-lobes; May to October. A garden escape now quite widely naturalised on rocks and walls.

BROOMRAPE FAMILY Orobranchaceae
Toothwort *Lathraea squamaria* is a short perennial that is wholly parasitic on tree-roots and has no green colour at all. The leaves are reduced to fleshy, creamy white scales, and the dull purple, tubular flowers are in a dense, one-sided spike; April to May. In damp woods, particularly underneath Hazel and Elms, scattered through most of the country, particularly in limestone areas.

Common Broomrape *Orobanche minor* is a short, purplish or yellowish annual, that parasitises the roots of other plants and has no green colour. The leaves are reduced to brownish scales, pressed to the stem, and the purplish brown, tubular flowers, 10–18 mm, are more or less 2-lipped; June to September. It grows on clovers and various hawkbits and allies, mainly in S England.

MOSCHATEL FAMILY Adoxaceae
Moschatel *Adoxa moschatellina* is a low perennial, forming bright green carpets of long-stalked trefoil leaves, with lobed leaflets. It produces small clusters of tiny, green flowers, 5–6 mm, 4 of them pointing at right angles to one another, the fifth straight up; March to May. Widespread in shady places, and commonest in the S and W; also on mountains.

VALERIAN FAMILY Valerianaceae have opposite, often pinnate leaves and clusters of small flowers with 5 joined petals.

Valerian *Valeriana officinalis* is a tall perennial with unequally pinnate leaves, the lowest on long stalks, the upper on short stalks or unstalked. It bears rounded heads of pale pink flowers, 4–6 mm; June to August. Common in damp grassland and woods, and in fens, and less frequently in dry grassland, when it is usually smaller.

Marsh Valerian *Valeriana dioica* is a short perennial with long, creeping runners bearing loose rosettes of small, untoothed, oval leaves, giving rise to flowering stems with unstalked, pinnate leaves. Male and female flowers are on different plants, the male 4–5 mm, the female 2–3 mm across; May to June. In wet meadows, fens, and mossy springs; scarce in the SE, rare in Scotland.

Red Valerian *Centranthus ruber* is a tall, rather bluish perennial with long, oval leaves, untoothed at the base of the stem, but with curved teeth near the top. The flowers, in dense clusters, are red, pink or white, 5 mm across and with a tube up to 10 mm long; May to September. Commonest on rocks and walls near the coast in SW England and Wales, scattered elsewhere.

Cornsalad *Valerianella locusta* is a short, rather delicate, branching annual with spoon-shaped lower leaves and oblong, untoothed upper leaves. The tiny pale lilac flowers, 1-2 mm, are in flat clusters surrounded by leafy bracts; April to August. Widespread in dry bare places, including sand-dunes, rocks and walls; scarce in Scotland.

HONEYSUCKLE FAMILY Caprifoliaceae
Twinflower *Linnaea borealis* was the famous botanist Linnaeus' favourite flower (he named it after himself). It has slender, creeping, woody stems, forming evergreen mats of oval leaves. The bell-shaped, pink flowers, 8 mm, are in nodding pairs on long stalks; June to August. A rare plant of Scottish pinewoods.

BUTTERWORT FAMILY Lentibulariaceae
Butterwort *Pinguicula vulgaris* (1) is a low perennial with a rosette of yellowish, insect-eating leaves, which are sticky and roll inwards to trap midges and other small insects. The violet, 2-lipped flowers, 10–15 mm, are on long stalks; May to August. Common on wet moors, bogs and mountain rocks in the N and W, rare elsewhere. **Pale Butterwort** *P. lusitanica* (2) has smaller, pale pinkish-lilac flowers. W coast.

PLANTAIN FAMILY Plantaginaceae have ribbed leaves in a rosette and dense spikes of tiny, 4-petalled flowers, coloured by the stamens.

Greater Plantain *Plantago major* has broad, oval leaves on long stalks, though in long grass they can be much narrower. The flower-spike is narrow and at least as long as its stalk, and has anthers that start purple and turn yellow; June to October. Extremely common on bare ground, particularly on paths and lawns; less frequent in grassland.

Hoary Plantain *Plantago media* is a hairy, often greyish perennial whose oval leaves taper gradually into a short stalk. The flower-spike is much shorter than in Greater Plantain, less than 10 cm long, and on a much longer stalk, up to 30 cm long. The stamens have purple filaments and conspicuous creamy anthers; May to August. Widespread in grassland on chalk and limestone from S Scotland southwards.

Buckshorn Plantain *Plantago coronopus* is a greyish, downy biennial with pinnately cut leaves. The flower-stalks are longer than the leaves and curve outwards and upwards from the rosette, ending in a short spike with conspicuous yellow anthers; May to October. A common plant of dry, sandy places and rocks, usually near the sea; occasional inland in the Midlands.

Sea Plantain *Plantago maritima* is a perennial with narrow, grass-like, fleshy, 3–5-veined leaves. It has a narrow, often pointed flower-spike on a stalk usually longer than the leaves; June to September. A characteristic salt-marsh plant, also found in turf affected by wind-blown sea-spray, and near streams in mountains.

Ribwort Plantain *Plantago lanceolata* is a perennial with long, narrow, oval leaves, prominently veined, and tapering into a long, winged stalk. It has a short, oval flower-spike, blackish brown with creamy anthers, on a very long, furrowed stalk; April to October. One of the commonest of grassland plants; in exposed positions (sea cliffs, mountains, etc) plants with much shorter flower-stalks occur.

ARROW-GRASS FAMILY Juncaginaceae **Sea Arrow-grass** *Triglochin maritima* (1) could be confused with Sea Plantain (above) but has flat, unveined leaves, which may be very long. It has 3-petalled flowers, in a long-stalked spike; May to September. A common salt-marsh plant. **Marsh Arrow-grass** *T. palustre* (2) has furrowed leaves and an arrow-shaped ripe fruit. Marshes and wet meadows, commonest in the N.

86

TEASEL FAMILY Dipsacaceae have opposite leaves and 4–5-petalled flowers in dense flat or rounded heads.

Field Scabious *Knautia arvensis* (1) is a short or tall, rather coarse and bristly perennial with a rosette of pinnately lobed leaves, and narrow, less lobed leaves on the stem. The flowers are in flat heads, 30–40 mm across, with enlarged petals around the rim; June to October. There are 2 rows of bracts under the flower-head. Common in dry, grassy places, except in NW Scotland. **Small Scabious** *Scabiosa columbaria* is more delicate, and has flower-heads 15–25 mm, with 1 row of bracts. Chalk and limestone grassland.

Devilsbit Scabious *Succisa pratensis* is a short, rarely tall perennial whose narrow, oval leaves often have dark blotches. It has deep blue–purple flowers in rounded heads, 15–25 mm across, surrounded by star-like bracts; June to October. Common in marshes and damp woods and grassland.

Teasel *Dipsacus fullonum* is a very tall biennial with lines of prickles on its angled stems. In its first year it produces a rosette of oblong leaves spotted with pale, pustulate prickles. The flowering stem bears a large, spiny, conical head, up to 80 cm long, with masses of reddish purple flowers; July to August. The dead heads survive through the winter. In bare and grassy places on heavy soils; common in the S, rarer north to S Scotland.

Small Teasel *Dipsacus pilosus* is like a small Teasel but the rosette leaves are oval and longer-stalked and the flower-heads are round, with white flowers and prominent violet anthers; July to September. Scattered in damp, usually shady places, mainly in central England.

Round-headed Rampion *Phyteuma tenerum* is a short perennial with a few long-stalked, narrow, oval leaves at the base of the stem, and 1–2 unstalked stem-leaves. It has slightly flattened, round heads of rich violet flowers; July to August. It can be distinguished from Scabiouses (above) by the long, narrow petal-lobes. Only on chalk grassland in S England, S of the Thames, but occasionally abundant.

Sheepsbit Scabious *Jasione montana* is a short biennial with narrow, oblong, wavy-edged leaves. The bright blue flowers are in a flattened, rounded head, up to 35 mm across; May to September. Common in the W on dry, grassy and heathy places on acid soils; near the sea dwarf forms are found.

1

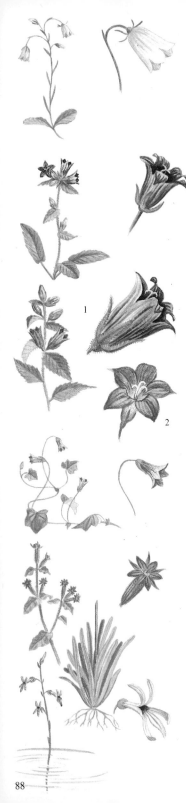

BELLFLOWER FAMILY Campanulaceae. Bellflowers all have 5-lobed, bell-shaped flowers.

Harebell *Campanula rotundifolia* is a delicate, short perennial with a few short-lived, rounded root-leaves, and very narrow, unstalked stem-leaves. The nodding, blue flowers, 15 mm long, are in a branching cluster; July to October. Common in dry, grassy and heathy places.

Clustered Bellflower *Campanula glomerata* is a low to short perennial with a stout, hairy stem and a loose cluster of long-stalked, narrowly heart-shaped leaves rising from the stem-base. The deep blue–violet flowers, 15–20 mm long, are in tight, upright, unstalked clusters; June to October. Frequent in grassland on chalk limestone in the S and E.

Nettle-leaved Bellflower *Campanula trachelium* (1) is a tall, roughly hairy perennial with angled stems. The lower leaves are heart-shaped and long-stalked, the upper oval, short-stalked, and toothed. The violet–blue flowers, 30–40 mm long, open first at the top of the leafy spikes; July to September. Woods and hedges in central and S England. **Spreading Bellflower** *C. patula* (2) is more delicate with narrower leaves and flatter flowers, only 15–25 mm; June to July. **Giant Bellflower** *C. latifolia* is bigger, and has larger, paler flowers (sometimes white) with narrower lobes, opening from the base of the spike. N England.

Ivy-leaved Bellflower *Wahlenbergia hederacea* is a delicate, creeping perennial with long-stalked, ivy-shaped leaves. The nodding, pale blue flowers, 6–10 mm long, are on very long thread-like stalks, rising from the leaf-junctions; July to August. A rather scarce plant of damp woods, heaths and moors in SW England and Wales.

Venus' Looking Glass *Legousia hybrida* is a low or short, upright annual with short, stiff hairs. It has wavy-edged, oblong, unstalked leaves, and small clusters of dull purple flowers, 8–15 mm across, which close up in dull weather; May to August. A locally common arable weed in S and E England.

Water Lobelia *Lobelia dortmanna* is a short perennial with runners connecting neat tufts of narrow, grass-like, submerged leaves, from which rise leafless stems bearing 2–6 nodding lilac flowers, 15–20 mm, clearly 2-lipped; July to August. Frequent in infertile lakes with clear water and stony bottoms in Scotland; scarce in the Lake District and Wales.

DAISY FAMILY Compositae. The composites are a very large family of plants, all of which have small flowers gathered into a flower-head which often looks like a single flower. There are 2 types of flowers: disc florets, in which the petal-tube ends in 5 short teeth; and ray florets, in which it has a strap-like projection. Composites may have heads with disc florets only (1, e.g. thistles, 2), ray florets only (3, e.g. dandelions, 4), or disc florets in the centre with ray florets round the edge (e.g. daisy, 5).

Hemp Agrimony *Eupatorium cannabinum* is a tall perennial with opposite pairs of deeply 3-lobed leaves, giving the appearance of a whorl of 6 narrow-toothed leaves. It has dense heads of rayless, dull pinkish mauve flowers; July to September. Common in the S, scattered in the N, and rare in Scotland, in damp woods, marshes, and by streams.

Canadian Goldenrod *Solidago canadensis* (6) is a tall perennial with unstalked, 3-veined leaves. The flower-heads have short rays and are densely crowded on to long, arching, 1-sided spikes in a branching cluster; July to October. A very common garden escape on waste ground. **Goldenrod** *S. virgaurea* (7) is shorter, sometimes low. It has broader leaves and loose, branched spikes of flower-heads. It is the native species and grows in woods, heaths, and on rocks, mainly in the N and W.

Daisy *Bellis perennis* is so familiar that it can hardly be confused. It is a low perennial with a neat rosette of spoon-shaped leaves and single, long-stalked flower-heads, 15–25 mm, with a yellow disc and white ray florets, usually red on their backs; all year. A very common plant of lawns and short turf, usually indicating low soil fertility.

Scentless Mayweed *Tripleurospermum inodorum* (8) is a short, weak-stemmed annual with feathery leaves, which are twice or thrice pinnate. It has daisy-like flower-heads, 15–40 mm, surrounded by brown-edged bracts; April to October. Widespread and common in disturbed ground. **Scented Mayweed** *Matricaria recutita* is pleasantly scented and more upright. The heads are 10–20 mm, with down-turned rays and green bracts. Not in Scotland.

Pineapple Mayweed *Matricaria matricarioides* emits a strong smell of pineapple when crushed. It is a low annual with feathery, twice pinnate leaves and greenish yellow flower-heads, 5–8 mm, which have no ray florets; June to October. An extremely common weed of disturbed ground, particularly paths and farmyards.

89

Sea Aster *Aster tripolium* (1) is a short or tall perennial with narrow, fleshy leaves and loose clusters of yellow flower-heads, which have yellow disc florets and blue-purple rays; July to October. The ray florets may be malformed or absent altogether. Common in salt-marshes and on coastal rocks. **Michaelmas Daisy** *A. novi-belgii* is tall and well-branched and not fleshy. A common garden escape, particularly by railway tracks.

Blue Fleabane *Erigeron acer* is a short annual or biennial with reddish stems, well-clothed in hairy, narrow, unstalked leaves. The flower-heads are 12–18 mm across, in loose clusters. They have yellow disc florets and upright, dull bluish purple rays; June to September. Widespread in dry, grassy places in S and E England, rare elsewhere.

Canadian Fleabane *Conyza canadensis* is a low, short or tall annual with stiff, branching stems bearing many small, narrow leaves; the rosette leaves die very early. It has small flower-heads, 3–5 mm, in open, branched spikes, each with yellow disc florets and whitish or dirty pale blue rays; July to October. Frequent in bare places in S and E England; rare elsewhere.

Marsh Cudweed *Gnaphalium uliginosum* (2) is a low, branching annual, covered with grey hairs. Its narrow leaves almost hide the tight clusters of flower-heads, 3–4 mm. The heads have yellowish disc florets, no rays, and brownish bracts; July to October. Widespread in damp, bare places. **Heath Cudweed** *G. sylvaticum* is a short, upright perennial, with long spikes of flower-heads with conspicuous brown bracts. Dry places.

Mountain Everlasting *Antennaria dioica* is a low, creeping perennial with long, leafy runners. The leaves are silvery white underneath and grey–green on top. It has separate male and female plants: the female plants have flower-heads 10–12 mm, with long, feathery bracts, while the male ones have heads 5–8 mm with rounded, spreading bracts resembling ray florets; June to July. Widespread in dry places in mountains; rare in England.

Ploughman's Spikenard *Inula conyza* is a short or tall perennial with purplish, downy stems and a rosette of broad leaves which are regularly mistaken for those of Foxglove (p.83). It has flattened clusters of rayless, yellow flower-heads, 8–10 mm, with purplish bracts; July to September. Widespread on dry chalk and limestone in S England.

Golden Samphire *Inula crithmoides* forms short or tall, bright green, fleshy clumps. It is a perennial with densely leafy stems and an open cluster of bright yellow, rayed heads, 20–25 mm; July to September. On sea-cliffs, shingle, and salt-marshes in S England; also in Wales and SW Scotland.

Fleabane *Pulicaria dysenterica* is a short, hairy perennial whose greyish stems have heart-shaped, wavy-edged leaves almost clasping them. The stems bear loose clusters of bright yellow, long-rayed flower-heads, 15–30 mm; July to September. Common in wet, grassy places in lowland England; rare in Scotland.

Trifid Bur-marigold *Bidens tripartita* (1) is a short annual with deeply 3-lobed leaves, the central lobe much the largest. It bears rather few rayless, dull yellow flower-heads, 15–25 mm; July to October. The fruits are bristly. Widespread on damp mud by ponds and streams in lowland England. **Nodding Bur Marigold** *B. cernua* has unlobed leaves and nodding heads.

Shaggy Soldier *Galinsoga ciliata* (2) is a short, hairy annual with opposite pairs of oval leaves. It has small flower-heads, 3–4 mm, with yellow disc florets and 4–5 flat, white, 3-lobed rays; June to October. Widespread as a weed of disturbed ground and greenhouses in SE and central England. **Gallant Soldier** *G. parviflora* is hairless and yellowish.

Yarrow *Achillea millefolium* has creeping runners and short, upright, woolly stems. The feathery leaves are dark green, twice or thrice pinnate, and narrow in outline. The stems end in flat-topped clusters of small flower-heads, 4–6 mm, with creamy disc florets and short, white or pink rays; June to November. Extremely common in grassy places.

Sneezewort *Achillea ptarmica* is a short perennial whose stems are hairy at the top, but not at the base. It has narrow, finely-toothed leaves, and loose clusters of white, short-rayed flower-heads, 12–18 mm; July to September. Common in damp grassland, particularly in the N and W.

Corn Marigold *Chrysanthemum segetum* is a short, rather bluish annual with narrow deeply-toothed leaves. The solitary flower heads, 35–65 mm, are a rich golden yellow and have long rays; June to October. A widespread weed on light sandy soils, occasionally appearing in very large numbers.

Tansy *Tanacetum vulgare* is a tall, downy aromatic perennial. The stems are well clothed in unstalked, pinnate leaves, which have pinnately-cut lobes. The rayless yellow flower-heads, 6–12 mm, are in dense, flat-topped clusters; July to October. Frequent by roadsides and in disturbed ground in lowland areas; formerly cultivated as a herb.

Feverfew *Tanacetum parthenium* is a short, strongly aromatic, downy perennial, with yellow–green, pinnate leaves. The flower heads have yellow disc florets and white rays, 10–25 mm across, and are in loose clusters; July to September. The rays are almost as wide as long. A widespread introduction on walls, disturbed ground, and roadsides; a double-flowered garden form sometimes escapes.

Ox-eye Daisy *Leucanthemum vulgare* is a short or tall perennial with a rosette of toothed leaves and a usually unbranched stem bearing a few dark green, pinnately-toothed, oblong leaves and a solitary daisy-like flower-head, 25–50 mm; May to September. Very common in dry, grassy places.

Butterbur *Petasites hybridus* produces short, stocky flowering spikes from March to May before the leaves appear. The flower-heads are lilac–pink and rayless, and male and female flowers are on separate plants. The leaves appear in summer, and are heart-shaped and up to 1 m across, forming huge patches in damp, grassy places and by streams; common everywhere except N Scotland.

Winter Heliotrope *Petasites fragrans* is a perennial with leaves present all the year round; they are heart-shaped and 10–20 cm across. The flowering stems appear at the same time as the new leaves in early spring (December to March), and are shorter than in Butterbur, with fewer, fragrant flower-heads. A widespread garden escape in damp places in S England, becoming scarce further N.

Mugwort *Artemisia vulgaris* (1) is a tall, downy perennial with pinnate leaves which are dark green on the top and silvery underneath. It has dense, branching spikes of small, brownish, rayless flower-heads, 2 mm across and 4 mm long; July to September. Common in waste places, less so in Scotland. **Wormwood** *A. absinthium* (2) is strongly aromatic and starts flowering in October.

Sea Wormwood *Artemisia maritima* (3) is a short, downy, greyish perennial with a strong aroma. It has pinnate leaves which are woolly on both sides and leafy, drooping spikes of small, yellow, rayless flower-heads, 1–2 mm; August to October. In the upper parts of salt-marshes from S Scotland southwards.

Coltsfoot *Tussilago farfara* forms large patches of heart-shaped leaves, 10–20 cm across, which appear after the flowers. The flower-heads are solitary, 15–35 mm, yellow and long-rayed, and appear on whitish stems with purple scales from February to April. Very common on bare ground. In fruit it forms dandelion-like 'clocks'.

Ragwort *Senecio jacobaea* (4) is a tall perennial with a ridged and branched stem, and pinnately-lobed leaves ending in a small, blunt lobe. The flower-heads are 15–25 mm in dense, flat-topped clusters; June to November. Very common in dry, particularly overgrazed grassland. **Hoary Ragwort** *S. erucifolius* is grey-downy and has narrow leaf-lobes, the end one pointed. Mainly in S and central England; not in Scotland.

Groundsel *Senecio vulgaris* (5) is a short annual with pinnately-lobed leaves with loose clusters of rayless flower-heads, though forms with yellow rays do occur; all year. Bracts have black tips. Very common in disturbed ground. **Sticky Groundsel** *S. viscosus* (6) has sticky hairs, green bracts, and yellow rays. **Heath Groundsel** *S. sylvaticus* is like Sticky Groundsel but has purple-tipped bracts. Dry, sandy soils.

Oxford Ragwort *Senecio squalidus* (7) is a short perennial, though usually short-lived. It has pinnately-lobed leaves, ending in a pointed lobe, and loose clusters of rayed flower-heads, 15–20 mm; April to November. The bracts are black-tipped. Increasingly common in walls and dry, bare places. **Marsh Ragwort** *S. aquaticus* has larger flower-heads, 25–30 mm, and green bracts. Damp grassland.

93

Lesser Burdock *Arctium minus* is a tall, thick-stemmed biennial with large, downy, heart-shaped leaves. It forms branching, bushy plants with arching, ridged stems, bearing many egg-shaped flower-heads, 15–30 mm; July to September. The heads have no rays but are surrounded by spiny bracts, which persist in fruit to form burs. Common in woods, hedges and waste places.

Carline Thistle *Carlina vulgaris* is a short biennial with an inconspicuous rosette of woolly leaves that soon dies. The flowering stems bear a few weakly spiny leaves and a few brown, rayless flower-heads, 20-40 mm, surrounded by long, narrow, golden brown bracts, which resemble ray florets; July to September, though flower-heads are visible much later. Common on chalk and limestone grassland in the S, rare in Scotland.

Musk Thistle *Carduus nutans* is a tall, usually unbranched biennial with spiny, pinnately-lobed stem-leaves and a winged, cottony stem. The large, rounded, rayless heads, 30–50 mm, usually borne singly, are always nodding; June to August. They are surrounded by a ring of spiny bracts. Common in bare and grassy places on chalk and limestone in most of England, but scarce in Wales and rare in NW England and Scotland.

Welted Thistle *Carduus acanthoides* is a tall biennial with a spiny-winged, cottony stem, except at the very top. The pinnately-lobed leaves have weak spines and cottony hairs underneath, and the flower-heads are 10–20 mm across, in tight clusters, surrounded by softly spiny bracts; June to September. Common in grassland in the S and E, scattered in the W and rare in the N.

Slender Thistle *Carduus tenuiflorus* is a slender, short or tall biennial which has branched stems with spiny wings to the very top. It has spiny leaves, cottony white underneath, and dense clusters of small, pinkish purple flower-heads, only 8–10 mm across; June to August. Widespread in grassy and disturbed places near the sea, except in NW Scotland.

Spear Thistle *Cirsium vulgare* is a tall biennial with extremely spiny, pinnately-lobed leaves, with a long end-lobe. The stem is ridged and cottony and has spiny wings between the leaves. The flower-heads are usually borne singly, 20–40 mm across, with a tuft of pale purple florets sitting on a spiny ball of bracts; July to September. Very common in disturbed ground and maltreated fields.

Melancholy Thistle *Cirsium helenioides* is a tall, unbranched perennial with a grooved, cottony, but unwinged stem. The lowest leaves are stalked, the rest unstalked, but all are narrow oval, toothed but scarcely spiny, and green above and white below. The flower-heads are solitary, 30–50 mm across, with an egg-shaped lower part which has broad, green, flattened bracts; June to August. In damp, grassy places in hills.

Meadow Thistle *Cirsium dissectum* is very similar to Melancholy Thistle (above) but is often shorter and its leaves are less thickly white-hairy underneath. It has smaller flower-heads, 20–25 mm across; June to July. Widespread but not common in damp, peaty grassland and fens in S England and Wales; it does not overlap with Melancholy Thistle.

Stemless Thistle *Cirsium acaulon* is the only thistle to bear its flower-heads at ground level, in the centre of a rosette of narrow, spiny, pinnately-lobed leaves. The bright purplish red flower-heads are 25–50 mm across; June to September. It is a perennial and the rosettes are a painful feature of short chalk and limestone turf in S England, N to Yorkshire.

Creeping Thistle *Cirsium arvense* is a tall perennial, creeping underground, and producing many upright, unwinged stems, which bear spiny, pinnately-lobed leaves. The leaves are sometimes hairy, the stem rarely so. The dull purplish lilac flower-heads, 15–25 mm, are borne in small clusters, and have purplish bracts; June to September. Very common in disturbed ground and an aggressive weed.

Woolly Thistle *Cirsium eriophorum* is a tall, sometimes very tall biennial which has huge, round, white-woolly flower-heads, topped by bright red–purple florets, 40–70 mm across; July to September. The leaves are unlike those of any other thistle – pinnate with narrow, spiny lobes, each divided into 2 leaflets pointing in different directions. Scattered in limestone grassland in England and Wales.

Marsh Thistle *Cirsium palustre* is a tall or very tall, but somewhat slender, biennial, with branching, spiny-winged, cottony stems. The leaves are pinnately-lobed and spiny, and the flower-heads are in dense, leafy clusters, each 10–15 mm, and either deep purplish red or dirty white; July to September. Very common in marshes and damp woods.

Black Knapweed *Centaurea nigra* (1) is a short perennial with tough, grooved stems and narrow leaves, untoothed at the top but toothed at the bottom of the stems. The stem is swollen beneath the solitary round flower-heads, 20–40 mm, which have black–brown bracts and red–purple florets; June to September. Very common in grassy places. *C. debauxii* has a slender, unswollen stem and the dark bracts have conspicuous pale bases. Commoner in the S.

Greater Knapweed *Centaurea scabiosa* is a short, often tall, downy, branching perennial with deeply-lobed leaves. It has large, reddish purple flower-heads, 30-50 mm across, with the outer row of disc florets expanded at the tip to give the appearance of ray florets; June to September. Common in dry, grassy places in the S and E, scarce in Wales and the N, rare in Scotland.

Cornflower *Centaurea cyanus* is a short or tall, usually well-branched annual, which has deeply-lobed lower leaves, gradually reducing up the stem to small, untoothed leaves at the top. Leaves and stem both bear thick, grey hairs. The flower-heads resemble those of Greater Knapweed (above), but are bright blue and only 15–30 mm; June to August. Once common, now a rare arable weed, mainly in E England.

Saw-wort *Serratula tinctoria* is a short or tall, rather delicate perennial, thistle-like but spineless, which has pinnate or deeply-lobed leaves with saw-toothed edges. Male and female flower-heads occur on separate plants, 15–20 mm across, contracted between the bracts and the florets into a flask shape; July to October. Common in the SW in damp, grassy places, often near woods; scattered elsewhere north to S Scotland.

Chicory *Cichorium intybus* is a tall, stiffly-branched perennial, whose lower leaves are stalked and lobed, while the upper are narrow, toothed, and clasp the stem. The flower-heads have ray florets only and are bright blue, 25–40 mm, in tight clusters; June to September. Common in the SE on bare and grassy chalk and limestone soils, becoming less common to the N and W.

Goatsbeard *Tragopogon pratensis* is a tall perennial with long, narrow, often twisted leaves, broadening at the base and forming a sheath round the stem. It bears large, solitary, yellow flower-heads on long stalks; May to August. The flower-heads only open on sunny mornings, and are partly hidden by long, green bracts. In fruit each forms a large 'dandelion clock'. Common in grassy places in lowland areas; scarce in Scotland.

Smooth Sowthistle *Sonchus oleraceus* (1) is a variable, short to very tall annual, whose hollow, angled, branching stems bear deeply pinnately-lobed leaves, with pointed, spreading lobes at the base. The stems exude milky juice if broken. The pale yellow flower-heads have green bracts, each 20–25 mm in a loose cluster; May to November. A common lowland weed. **Prickly Sowthistle** *S. asper* has less lobed leaves, spiny-edged, with rounded bases clasping the stem.

Perennial Sowthistle *Sonchus arvensis* (2) is a tall or very tall perennial whose pinnately-lobed, greyish leaves are softly spiny-edged and clasp the stem with rounded lobes. The flower-heads are golden, 40–50 mm, in branching clusters; the bracts are covered with sticky, yellowish hairs; July to September. It forms clumps in damp, grassy places, by streams, salt-marshes and on bare ground.

Nipplewort *Lapsana communis* (3) is a tall, tough-stemmed, branching annual. The lower leaves have a large end lobe and few or no side lobes, while the upper ones are simply oval and toothed. The pale yellow flower-heads, 15–20 mm, are in branching clusters; June to October. The fruits are pale greenish brown and hairless. Common except in N Scotland in open woods and hedges, on roadsides and bare ground.

Dandelion *Taraxacum* Section *Vulgaria* (4) is the commonest group of 'micro-species' in this enormous genus. It has a rosette of deeply-lobed leaves and large, yellow flower-heads, 35–50 mm, solitary on long, hollow stalks which contain a milky juice; all year but especially in spring. The bracts are often turned out or down. The fruits form the familiar clock. Very common in bare and grassy places. **Lesser Dandelion** *T.* Section *Erythrosperma* is more delicate, with flower-heads 15–25 mm. Dry places.

Few-leaved Hawkweed *Hieracium murorum* (5) is the commonest of this large genus. It is a tall, hairy perennial with a rosette of toothed leaves and an almost leafless flowering stem. The pale yellow flower-heads, 20–30 mm, are borne in loose clusters; June to August. The stems exude a milky juice. Common in dry, grassy and rocky places and on walls.

Mouse-ear Hawkweed *Hieracium pilosella* (6) is a low, creeping perennial with long runners and tight rosettes of elliptical leaves which bear long, white hairs. The lemon-yellow flower-heads, 20–30 mm, are borne singly on leafless stalks, the outer florets reddish underneath; May to October. Dry, bare places. **Orange Hawkweed** *H. auranti-acum* (7) has small clusters of orange flower-heads.

97

Wall Lettuce *Mycelis muralis* is a slender, short or tall, branching perennial with raggedly pinnately-lobed leaves. The leaves are all thin and often reddish, sometimes almost upright, and the upper ones clasp the stem. The flower-heads are in loose clusters, each 7–10 mm and with only 5 florets; July to September. Widespread in dry, often shady places, scarce in E and SW England and Scotland.

Bristly Oxtongue *Picris echioides* (1) is a tall, roughly hairy, branching perennial with a thick, ridged stem bearing narrow, wavy-edged leaves, which have white spots on them. The pale yellow flower-heads, 20–25 mm, are surrounded by small inner and a few large triangular outer bracts; June to October. Bare and grassy places, mainly in S England. **Hawkweed Oxtongue** *P. hieracioides* has narrower, toothed leaves and narrow, spreading, black-hairy bracts.

Rough Hawkbit *Leontodon hispidus* is a short and very hairy perennial with a rosette of shallowly-lobed leaves which bear long, forked hairs. Each rosette usually produces a single flower-head on a leafless stalk. The heads are 25–40 mm, rich golden yellow, with very dark green, hairy bracts; June to October. Common in grassland, often on limestone, in England; scarce in Scotland.

Catsear *Hypochaeris radicata* is a short perennial with a rosette of hairy, lobed leaves. The flowers are borne on a short to tall, branching shoot, which bears a few flattened, minute scales and is swollen beneath the heads. Flower-heads are 25–40 mm across and bright yellow; June to September. Very common in dry grassland.

Smooth Hawksbeard *Crepis capillaris* is a tall, almost hairless perennial with deeply-lobed leaves in a basal rosette and arrow-shaped leaves clasping the flowering stems. The branched stems bear loose clusters of yellow flower-heads, 10–15 mm, which are reddish in bud; June to October. Very common in dry, grassy places, except in mountains.

Beaked Hawksbeard *Crepis vesicaria* (2) is a tall, softly hairy perennial, which has pinnately-lobed leaves clasping the stems with pointed lobes. The flower-heads, 15–25 mm, are in many-flowered clusters and have spreading bracts; May to July. Common on dry ground in S and E England. **Marsh Hawksbeard** *C. paludosa* has narrow, toothed leaves, and few-flowered clusters of heads with black-hairy bracts. Marshy places in Scotland and N England.

MONOCOTYLEDONS are plants which are almost never woody (except Butcher's Broom, p.102), have usually parallel-sided and veined leaves, and have their petals and sepals in threes or sixes. Their name derives from their possessing only one seed-leaf or cotyledon on germination.

monocotyledon

dicotyledon

WATER-PLANTAIN FAMILY
Alismataceae

Common Water-plantain *Alisma plantago-aquatica* is a tall, hairless perennial with oval leaves which have prominent parallel veins. The flowers are in branching whorls, each 8–10 mm, pale lilac and only open in the afternoon; June to September. Very common on mud beside or in fresh water, becoming scarce in Scotland.

Arrowhead *Sagittaria sagittifolia* is a tall aquatic perennial with 3 sorts of leaves. The submerged leaves are narrow and ribbon-like, while the floating leaves are narrowly oval, and some leaves project from the water on long stalks and are arrow-shaped. The 3-petalled flowers, 20 mm, have purple centres; July to August. Widespread in still and slow-flowing water in S England; not in Scotland.

FLOWERING RUSH FAMILY
Butomaceae

Flowering Rush *Butomus umbellatus* is a tall or very tall perennial with long, narrow, rush-like, 3-cornered leaves, and rounded flowering stems bearing a cluster of pink flowers, 25–30 mm; July to August. Each flower-stalk rises from the same point on the stem but they vary in length. Widespread at the edge of fresh water, commonest in E and central England; only as an introduction in Scotland.

FROGBIT FAMILY Hydrocharitaceae
Frogbit *Hydrocharis morsus-ranae* has long, creeping, submerged runners from which rise groups of glossy dark green, or sometimes coppery, leaves, which are rounded or kidney-shaped, and clusters of 2 or 3 3-petalled white flowers, 20 mm, with a yellow spot in the centre; July to August. A rather uncommon plant of shallow, still, fresh water in central and S England.

Water Soldier *Stratiotes aloides* is an extraordinary plant, hidden under water for most of the year. In summer large rosettes of long, spiny-toothed leaves rise to the surface and bear solitary white female flowers, 30–40 mm (male plants are not known in Britain). The leaves resemble those of an Aloe, hence the name *aloides*. A rare native of base-rich water in E and central England, but often planted.

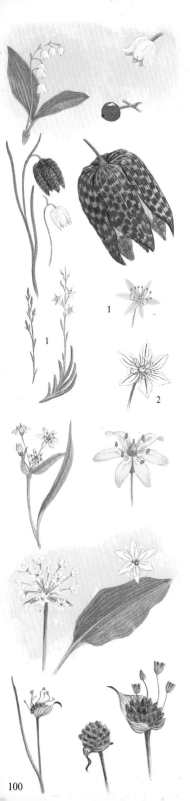

LILY FAMILY Liliaceae have 3 petals and 3 sepals, often similar and so apparently 6-petalled. Mainly bulbous.

Lily of the Valley *Convallaria majalis* is a short, creeping perennial, forming large patches of stems, each with 2 shoots, one ending in a pair of dark green, broadly elliptical leaves, and the other in a 1-sided spike of bell-shaped, creamy, fragrant flowers, May to June. Its fruit is a round, red berry. Scattered throughout England and Wales in dry woods, usually on chalk or limestone; rare in Scotland.

Fritillary *Fritillaria meleagris* is a short perennial, inconspicuous when not in flower with its grass-like leaves, but unmistakable in April and May, when the stems bear large, nodding, bell-shaped flowers, 30–50 mm long, which may be dull purple (typically) to creamy white, regularly chequered with darker patches. A fast declining species of unploughed, damp lowland meadows in S and central England.

Bog Asphodel *Narthecium ossifragum* (1) is a short perennial with stiff tufts of iris-like, curved leaves, from which rise leafless stems bearing spikes of bright orange–yellow flowers; July to August. In fruit the stem and capsules turn vermilion. Peaty areas in mountains, also in lowland wet heaths; not in central and E England. **Snowdon Lily** *Lloydia serotina* (2) has larger flowers, white with purple veins. Only in Snowdonia.

Yellow Star of Bethlehem *Gagea lutea* is a delicate, low, bulbous perennial. Each plant bears a single, flattened, yellowish leaf, prominently ridged on the back, and a short, leafless stem ending in 2 leaf-like bracts and a small cluster of yellow–green, star-like flowers, 20 mm, each with a green band on the underside of the petals; April to May. Widely scattered in damp woods.

Ramsons *Allium ursinum* exudes a powerful smell of garlic. It bears broadly elliptical leaves and leafless flowering stems, ending in a round cluster of star-like, white flowers, 15–20 mm; April to June. It forms clumps, sometimes large, in damp woods.

Crow Garlic *Allium vineale* is a short or tall perennial with hollow and very narrow leaves, folded almost to form a cylinder. The flowers are in a tight, round head, half-hidden by a papery bract. The heads normally contain a few long-stalked, pink or greenish flowers and egg-shaped, green bulbils; sometimes there are no flowers. Widespread on roadsides, dry grassland, and cultivated land, north to S Scotland.

Bluebell *Hyacinthoides non-scriptus* (1) is a very familiar short, bulbous perennial, which produces clumps of narrow, keeled leaves and a single, long, leafless, 1-sided spike of bright-blue bell-shaped flowers, 15–20 mm, with creamy anthers; April to June. Common in woods, often forming huge colonies on well-drained soils; also in the open, in upland grassland and on sea-cliffs. **Spanish Bluebell** *H. hispanicus* is the garden species, with an upright spike of larger, blue-anthered flowers.

Spring Squill *Scilla verna* is a low perennial which produces a few curly, slightly fleshy, narrow, dark green leaves, followed by a rather squat spike of pale violet–blue, star-like flowers, 10–15 mm; April to June. The black seeds are conspicuous in July in coastal grassland in Wales and SW England, and here and there in W and N Scotland and the Scottish islands.

Solomon's Seal *Polygonatum multiflorum* is a tall perennial with rounded, arching stems, bearing alternating, elliptical leaves which rise away from the stem, and hanging clusters of 1–3 bell-shaped flowers, 10–15 mm; May to June. The fruit is a black berry. In woods and scrub, mainly in S England, but scattered elsewhere. The garden plant *P.* x *hybridum*, with larger flowers, is widely naturalised.

Meadow Saffron *Colchicum autumnale* is a short, bulbous perennial producing pale rosy mauve crocus-like flowers on long, white stalks in autumn. The flowers wither and the fruit matures underground and appears in spring along with the leaves The flowers have 6 stamens. A scarce plant of damp woods and meadows, scattered from the Lake District southwards.

Herb Paris *Paris quadrifolia* is a short perennial with a naked stem topped by a whorl of 4 unstalked, oval leaves and a single flower. The flower has 4 long, narrow, green sepals, but the petals are thread-like and inconspicuous; May to June. The fruit is a single black berry. Scattered in woods, most often on limestone, and commonest in S England.

Common Star of Bethlehem *Ornithogalum umbellatum* is a short perennial whose narrow, floppy leaves each have a conspicuous white stripe down the centre. The star-like flowers, 25–40 mm, have white petals with a green stripe on the back, and are in loose clusters; May to June. Widespread in dry, grassy places; commonest in the S and E but more rarely north to NE Scotland.

Butcher's Broom *Ruscus aculeatus* is an ever-green bush, less than 1 m high, which has stiff, ridged stems. It has no real leaves, but instead the branches are flattened to resemble oval leaves with a sharp, spiny tip, though their origin is betrayed by their bearing 1–2 tiny, 6-petalled flowers, 2–3 mm, on the upper surface; January to April. The berry is almost as large as the 'leaf'. Shady places in S England.

DAFFODIL FAMILY   Amaryllidaceae
Wild Daffodil *Narcissus pseudonarcissus* is a short, bulbous perennial producing a tuft of parallel-sided, greyish leaves and a typical daffodil flower on a flattened stalk. It can be told from cultivated forms by the darker yellow 'trumpet', which is as long as the outer ring of petals. Flowers are up to 60 mm; March to April. Native in woods and grassland here and there in England and Wales.

Snowdrop *Galanthus nivalis* is the commonest planted species of snowdrop, with its narrow, greyish leaves and nodding white flowers. The outer ring of petals is pure white, while the inner ones are notched and green-tipped; January to March. Widespread in damp woods and hedgebanks and by streams. It may establish itself from an odd bulb thrown away in garden rubbish.

IRIS FAMILY   Iridaceae
Yellow Iris *Iris pseudacorus* is a tall, branching perennial with flat, often wrinkled, sword-shaped leaves rising from the creeping underground stem. The flowers are like a garden iris but yellow, 80–100 mm across; June to August. The capsules split in October to reveal flat brown seeds. Wet places.

Stinking Iris *Iris foetidissima* is a tall perennial, shorter than Yellow Iris, which has dark green leaves with an unpleasant smell if crushed. The iris-like flowers are 70–80 mm; May to July. The seed capsules open right out to reveal the bright orange seeds. Widespread in scrub and hedges, particularly near the sea, in S England; introduced north to S Scotland.

ARUM FAMILY   Araceae
Lords and Ladies *Arum maculatum* is a short perennial whose conspicuous, dark green, arrow-shaped, often dark-spotted leaves appear very early in the year. The flowers, in the base of the pitcher-like 'spathe', are indicated to pollinating flies by the finger-like purple 'spadix'; April to May. The spathe withers in fruit, revealing orange-red berries. Common in woods north to S Scotland.

ORCHID FAMILY Orchidaceae are perennials, and hairless except for some helleborines (p.105). They have simple leaves and 2-lipped, often spurred flowers.

Bee Orchid *Ophrys apifera* is short, with a rosette of elliptical leaves, and 1–2 stem-leaves, pressed against or wrapped round the stem. The flowers have pink sepals and green upper petals, with an expanded lower lip like a bumblebee; June to July. Widespread in S, E, and central England in chalk and limestone grassland; only on dunes in the W.

Fly Orchid *Ophrys insectifera* is more delicate and sometimes taller than Bee Orchid, and it has narrower, somewhat glossy leaves. The flowers have yellow–green sepals, 2 very thin, brown petals, and a 3-lobed lower lip, 10–12 mm long, whose central lobe is much the longest, notched, and with a bluish patch at the top; May to June. Scattered from N England southwards in woods and dry grassland on chalk and limestone.

Early Purple Orchid *Orchis mascula* (1) has a rosette of pointed, oblong leaves, usually blotched with blackish spots, and a short or tall stem ending in a dense spike of rich purple flowers, occasionally paler or even almost white; April to June – the first orchid to flower. The lower lip is 3-lobed, 8–12 mm, and there is a long spur, about 15 mm. Widespread, sometimes common, particularly on limestone, in woods and grassland.
Green Winged Orchid *O. morio* (2) has narrow oblong, pointed, unspotted leaves and fragrant purple, pink or white flowers, May to June. Grassland.

Burnt Orchid *Orchis ustulata* is one of the smallest orchids, with unspotted leaves and a dense spike of flowers which are deep maroon in bud (at the top) but open almost white, giving the spike a burnt appearance; May to June. Each flower has a maroon hood and a 4-lobed, pink-dotted, white lip. Scattered in dry, grazed chalk and limestone grassland throughout England.

Twayblade *Listera ovata* has a single short or tall stem, with a pair of broad, oval leaves near its base, and a long, loose spike of yellow–green, unspurred flowers, which have a long, forked lower lip; May to July. One of the commonest orchids, growing in many soil types in woods and grassland.

Man Orchid *Aceras anthropophorum* has a rosette of narrowly oblong, shiny leaves, and a short stem with a dense spike of manikin-like flowers. The sepals and upper petals form a hood and the lower lip is 4-lobed, resembling arms and legs; May to June. Uncommon on chalk and limestone grassland in S, E, and central England.

103

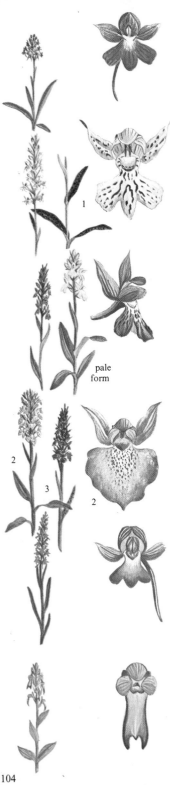

Pyramidal Orchid *Anacamptis pyramidalis* has narrow, unspotted leaves, often pressed tightly against the stem, and a short stem bearing a dense, pyramidal spike of rosy purple flowers. The sepals and petals are similarly coloured, and the lip is 3-lobed, 5–6 mm long; June to August. Each flower has a long, narrow spur. Widespread in dry chalk and limestone grassland in S and E England, mainly on sand-dunes in the W.

Common Spotted Orchid *Dactylorhiza fuchsii* (1) is the commonest orchid in Britain. It has rather narrow, dark-spotted leaves on the short stems, and a more or less dense spike of pink or rarely white flowers, marked with dull purple. The 3-lobed lip is about 10 mm across; June to August. Grassy places, mainly on chalk and limestone. **Heath Spotted Orchid** *D. maculata* grows on wetter, acid soils; it has a toothed lip.

Early Marsh Orchid *Dactylorhiza incarnata* has rather narrow, yellowish green leaves, which are often folded about the middle and at the tip. The flowers range from salmon-pink to dirty dark red and creamy yellow, but the lower lip always has 2 side lobes folded right back; May to June. Widespread, though never common, in damp grassland, fens and dune-slacks.

Southern Marsh Orchid *Dactylorhiza praetermissa* (2) has dark green leaves and pinkish purple flowers, with a broad, flat lip and a short, fat spur; June to July. In marshes, fens, and dune-slacks, widespread in the S, scattered north to N England. **Northern Marsh Orchid** *D. purpurella* (3) has a complementary distribution – in Scotland, N England and N Wales. The deep purple flowers have a narrow lip and tapering spur.

Fragrant Orchid *Gymnadenia conopsea* has narrow leaves on the short stems and rather dense spikes of very fragrant, pale purplish pink flowers, distinguished by the spreading pink sepals, the 3-lobed lip, and the very long, fine spur; June to July. In damp and dry grassland, though preferring chalk and limestone soils; rare in the Midlands.

Frog Orchid *Coeloglossum viride* is often very small and inconspicuous. It has narrow, oval leaves, in a rosette and up the stem, and a loose spike of yellowish flowers tinged with orange–brown. The sepals and petals form a hood and the lip is strap-shaped and notched, with a small tooth in the notch; June to August. Chalk grassland, dune-slacks, and mountain grassland; widespread but very scattered.

104

Greater Butterfly Orchid *Platanthera chlorantha* (1) has a single pair of broad, glossy, ribbed leaves at the base of the short or tall stem, and a few small leaves on the stem. The flowers are greenish with a long, narrow, undivided lip and a very long spur, and the pollen-sacs making an inverted V; June to July. Woods, heaths, and grassland, scattered but scarce in the E. **Lesser Butterfly Orchid** *P. bifolia* (2) is smaller and has parallel pollen-sacs. Moorland.

White Helleborine *Cephalanthera damasonium* is short, with broad leaves becoming narrower up the stem and merging into leaf-like bracts in the flower-spike. The creamy white flowers appear almost tubular as the petals never spread fully open to reveal the yellow-spotted lip; May to July. Mainly in beechwoods on chalk in S England.

Marsh Helleborine *Epipactis palustris* is short and downy, and has several prominently veined, folded, often purple-tinged leaves. The flowers have pinkish or purplish brown sepals, crimson-streaked upper petals, and the lower petal with a frilly lip; July to August. In fens and dune-slacks, mainly in the S but scattered north to S Scotland.

Broad-leaved Helleborine *Epipactis helleborine* has tall, downy, often purple-tinged stems, with a spiral of broad, sometimes almost circular, ribbed leaves. The flowers are in a rather 1-sided spike, and may vary from green to purple, but the sepals are usually greener than the petals; July to August. The lower lip has a point turned underneath. Mainly in woods; common in the S, more rarely N to central Scotland.

Autumn Lady's Tresses *Spiranthes spiralis* has a rosette of pointed, oval, bluish green leaves that wither before the low, almost leafless flowering stems appear in August and September. The tiny, unspurred, white flowers are arranged in a single row in a spiral around the spike. Dry grassland in the S, rarely N to Yorkshire; it often reappears at a site after a long absence.

Birdsnest Orchid *Neottia nidus-avis* is an extraordinary plant that actually parasitises fungi in the soil, for it has no chlorophyll. It can be told from broomrapes (p.84) by the 2-lobed lip, and by its habitat – it grows in woods, often of beech, in deep shade and frequently where there are no other plants. Flowering spikes appear from May to July.

105

RARE ORCHIDS

Lady's Slipper *Cypripedium calceolus* has much the largest flower of any British orchid. It was once widespread in the limestone hills of N England, but has been reduced by collectors to the point of extinction. It has broad, ribbed leaves and the flowers have dull purple sepals and petals, apart from the huge yellow, slipper-like lip; May to June.

Red Helleborine *Cephalanthera rubra* is very similar to White Helleborine (p.105), but has much narrower leaves and bright rose-purple flowers; June to July. It grows in beechwoods, often in deep shade, in one or two places on the chalk in S England.

Lizard Orchid *Himantoglossum hircinum* is one of the most easily recognised orchids with its olive-grey flowers, streaked with purple, which have an immensely long, strap-like lip, with 2 shorter side lobes; June to July. The flowers are in long spikes, with the lizard's 'tails' often becoming entangled, and they smell of goats. A rare plant of tall grassland, scrub, and wood-edges in S and E England.

Lady Orchid *Orchis purpurea* has tall, stout stems rising from a rosette of broad, glossy, unspotted leaves. The flowers are in the shape of a manikin, with a dull purple hood or 'head' and a 4-lobed lip, resembling arms and legs; May to June. It can be told from similar-sized species by the contrast between dark hood and pale lip. It grows in beechwoods and scrub on chalk in a few places in SE England.

Monkey Orchid *Orchis simia* is a short plant with a rosette of leaves and a few others on the stem. It has 'monkey'-shaped flowers, which have a hood of sepals and petals, and a long lip, 4-lobed to resemble arms and legs, but with long, narrow lobes, and with a tooth between the 'legs', giving the monkey a 'tail'; May to June. On chalk grassland in one or two places in S and E England.

Military Orchid *Orchis militaris* is very similar to Monkey Orchid, though larger. The lobes on the lip are shorter and broader, and the flowers are often darker purple with dark markings; May to June. Extremely rare in woods and scrub on chalk in S and E England, though occasionally in some quantity.

WATERWEEDS are an ecological grouping, drawn from several very different families, but brought together here because the problems of growing under water have resulted in their evolving a similar appearance. In addition they tend to have inconspicuous flowers. Many waterweeds have feathery leaves to increase their surface area for 'breathing' and photosynthesis – see also Water Crowfoot (p.39) and Water Violet (p.68).

Spiked Water Milfoil *Myriophyllum spicatum* (1) has long, submerged stems bearing whorls of 4 pinnate leaves with very fine leaflets. The stems end in emergent flower-spikes, with whorls of 4 tiny flowers; June to July. Widespread in still, fresh water, commonest in the E. **Whorled Water Milfoil** *M. verticillatum* has leaves mainly in whorls of 5, and leafy flower-spikes. Mainly in E England. *M. alterniflorum* has smaller leaves and flowers alternating on the spike. Commonest in Scotland.

Marestail *Hippuris vulgaris* (2) has stout stems bearing whorls of 6–12 flat, narrow, undivided leaves, some submerged, some out of the water. The minute greenish flowers are at the base of the aerial leaves. Distinguished from horsetails (p.121) by its flat leaves and by having flowers. Widespread in slow or still fresh water, but never common; rare in the W.

Water Starwort *Callitriche stagnalis* (3) is the commonest of a very confusing genus, which are only reliably distinguished by their fruits. It has long, trailing, floating stems bearing pairs of elliptical or oval leaves, often ending in a floating rosette. The tiny flowers are at the base of these rosette leaves; May to September. Common in slow or still fresh water and on mud.

Greater Bladderwort *Utricularia vulgaris* (4) is an extraordinary aquatic carnivorous plant with very finely divided leaves which bear small bladders; these act like vacuum cleaners, sucking in animals which trigger the mechanism. The 2-lipped yellow flowers appear on long spikes in July and August. Widespread but local in still, fresh, often peaty water.

Floating Bur-reed *Sparganium angustifolium* (5) differs from Branched Bur-reed (p.111) in having long, floating stems with long, narrow leaves, and a simple unbranched flowering spike. The flowers are in round, spiky balls. In peaty water in the hills of the N and W.

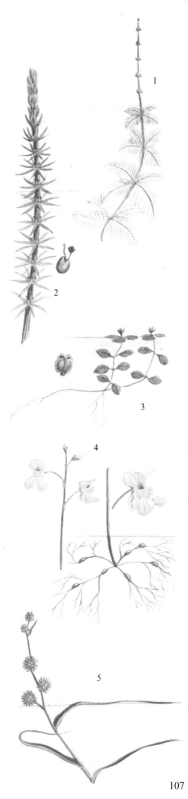

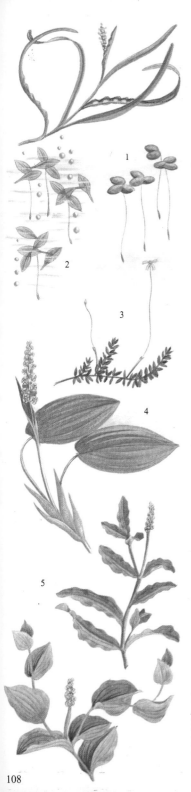

Eel-grass *Zostera marina* is one of the very few higher plants which actually grows in the sea, below the low-water mark of the tides, and so often escapes attention. It has very long, narrow leaves and short, dense spikes of minute green flowers; June to August. It sometimes forms immense beds on muddy shores, and is an important source of food for birds.

Common Duckweed *Lemna minor* (1) is a minute floating plant, round and flat, multiplying by division, so sometimes 2 or 3 'siblings' may be joined. Flowers (when produced) are almost invisible. Very common on still water, except in N Scotland. **Ivy-leaved Duckweed** *L. trisulca* (2) tapers and branches to give an ivy-leaf shape. It floats just under the surface. Scarce in the N and W.

Canadian Pondweed *Elodea canadensis* (3) is a plant introduced from N America over 100 years ago, which became very common and is now apparently decreasing. It has whorls of leaves in threes and sometimes floating flowers on long stalks. In ditches and slow streams, except in N Scotland. It is apparently being replaced by the similar **Nuttall's Pondweed** *E. nuttalli*, which has much narrower leaves.

Broad-leaved Pondweed *Potamogeton natans* (4) has floating, oval leaves, sometimes on very long stems. The floating leaves have a joint between them and the stem. Underwater leaves are poorly developed and very narrow. The flowers are in short, dense spikes, appearing above water from May to September. Common in fertile, still and slow-flowing water. **Bow Pondweed** *P. polygonifolius* has both submerged and floating leaves, the latter unjointed. In shallow, peaty water, mainly in the N and W.

Curled Pondweed *Potamogeton crispus* (5) is entirely submerged, except for the flower-spikes, and easily overlooked. It has wavy-edged leaves, about 5 times as long as broad. The fruits have a long, curved beak. Common in still or slow fresh water, except in N Scotland. **Small Pondweed** *P. berchtoldii* has even narrower, smooth, 3-veined leaves, and **Fennel-leaved Pondweed** *P. pectinatus* has thread-like leaves; it is water-pollinated.

Perfoliate Pondweed *Potamogeton perfoliatus* is also entirely submerged, except for its flower-spike. It has broad, unstalked, heart-shaped leaves, clasping the stems. Common in still and slow fresh water in the Midlands, but absent from mountains.

Rushes, Sedges and Grasses

Rushes, sedges and grasses are often overlooked by naturalists and they have gained an unnecessarily fierce reputation for difficulty. It is relatively easy to identify the commoner species in all 3 groups, once the nature of their specialised flowers is understood.

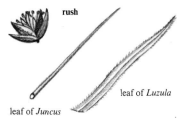

rush

leaf of *Juncus*

leaf of *Luzula*

Rushes are the most distinct. They have 'normal' flowers which resemble those of the Lily Family, with very small petals. There are 2 genera of rushes – the ordinary rushes *Juncus*, with cylindrical or deeply channelled, hairless leaves, and the Wood-rushes *Luzula*, with flat, hairy, grass-like leaves.

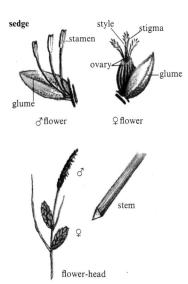

sedge

stamen

style

stigma

ovary

glume

glume

♂ flower

♀ flower

♂

♀

stem

flower-head

Sedges have more complicated flowers. Each is at the base of a small bract called a glume, and the flower may be male or female. The male flowers have 3 stamens and the female flowers 2 or 3 styles – this is a useful character. Nearly all sedges have obviously 3-angled stems. Most sedges have several flower-heads on a spike, and these may be all male, all female, or a mixture. In many sedges the male flowers are different from the female ones and the male spikes may be distinct from the female.

The Sedge Family contains a number of other genera than the true sedges *Carex*, most of which are easy to distinguish. Sedges and their allies are characteristic of very infertile soils.

Grasses have the most complex flowers of all. They are arranged in 'spikelets' and gathered together into a flower-head, which may be tight and unstalked as in Timothy or Ryegrass, or loose and stalked as in fescues and meadow grasses. Each spikelet consists of 2 bracts called 'glumes', enclosing one or more flowers, each of which comprises 2 'petals' known as the lemma and the palea, inside which are the

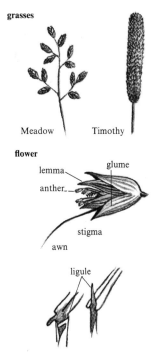

grasses

Meadow

Timothy

flower

lemma

glume

anther

stigma

awn

ligule

stamens and styles. The length of the glumes and the structure of the lemma (it often bears a long bristle called an awn) are important characters.

When identifying grasses and sedges, one further character is useful: where the leaves join the stem, a tiny membrane often wraps around the stem. This is called the ligule; its presence, length and shape are worth noting.

RUSH FAMILY   Juncaceae

Heath Rush *Juncus squarrosus* (1) is a short, tufted perennial with stiff, channelled, deep green leaves, forming a wiry rosette. The flowers are in small, tight clusters on a very stiff stem, and the rich brown petals are slightly longer than the rounded seed capsule; June to August. Very common on moorland on peaty soils, particularly in the W.

Hard Rush *Juncus inflexus* (2) is a tall perennial which appears to have long, green, leafless stems, each bearing a branching cluster of flowers about two-thirds of the way up (in fact the top part of the stem is a bract). The stems are bluish green and have conspicuous ridges. Flowers appear from June to August and ripen to a pointed, dark brown seed capsule. Common on damp soils.

Soft Rush *Juncus effusus* (3) differs from Hard Rush in having greener stems which bear very many fine ridges, and so seem smooth and glossy. The flowers may either be in a loose cluster or in a compact bunch (known as variety *compressus*; 4); the seed capsule is blunt and greenish. It is commoner on acid soils. **Clustered Rush** *J. conglomeratus* (5) differs from *J. effusus* var. *compressus* in having the stem more ridged just beneath the flowers, and a pointed capsule.

Jointed Rush *Juncus articulatus* (6) is a short or tall perennial whose rounded stems bear several flattened cylindrical, curving leaves, with internal divisions which can be felt by running a finger-nail up and down them. The flowers are in widely branching clusters; June to September. Common in wet, grassy places. **Sharp-flowered Rush** *J. acutiflorus* has unflattened leaves with more conspicuous divisions.

Great Woodrush *Luzula sylvatica* (7) is a tall, spreading perennial forming large clumps of long, wide, grass-like leaves, edged with long, silky hairs. The stems bear shorter leaves and end in a widely branching cluster of reddish brown flowers; May to July. Common in wet woods, particularly on banks by streams, mainly in the N and W.

Field Woodrush *Luzula campestris* (8) is a low, tufted perennial whose flat leaves have scattered, but long and conspicuous, hairs. The flowers are in tight groups in a loose cluster; March to June. The anthers are very long. Very common in short grassland. **Heath Woodrush** *L. multiflora* is short, with more widely spaced flowers which have shorter anthers; April to June. Heaths.

SEDGE FAMILY Cyperaceae
AND ALLIES —

Branched Bur-reed *Sparganium erectum* (1) is not in fact a member of the Sedge Family, but is closely related. It has long, bright green leaves, 3-cornered like the sedges, but very distinctive spherical flower-heads in long, branching spikes; June to August. Common in still and slow fresh water. **Unbranched Bur-reed** *S. simplex* has unbranched flower-spikes.

Great Reedmace *Typha latifolia* (2) is another close relative of the sedges. It has long, greyish leaves, up to about 20 mm wide, which have a cellular interior structure and so a springy feel when squeezed. The flower-spikes are unmistakable – the sausage contains the female flowers, the long, thin spike the male; June to August. Very common on the edge of fresh water. **Lesser Reedmace** *T. angustifolia* has narrower leaves (to 5 mm) and a smaller spike with the male flowers separated from the female. Less common.

Common Cotton-grass *Eriophorum angustifolium* (3) is a short, creeping perennial which has slightly folded leaves and a small group of drooping flower-spikes; May to June. It derives its name from the long, cottony hairs attached to the fruits; July. Common and forming large patches which are very conspicuous in fruit, in wet, peaty places, and so most frequent in the N and W. **Hare's-tail Cotton-grass** *E. vaginatum* (4) differs in having conspicuous cylindrical sheaths where the leaves join the stem, and in having a single upright flower-spike.

Common Spike-rush *Eleocharis palustris* (5) is a short, creeping and tufted perennial with round, green, leafless stems ending in tight, egg-shaped spikes, 5–20 mm long; May to July. Common in wet, grassy places. **Deer-grass** *Scirpus caespitosus* is more tufted and has a small leaf just below the smaller flower-spike. On damp heaths and bogs, particularly in the N and W.

Black Bog-rush *Schoenus nigricans* (6) is a short or tall, tufted perennial with long leaves rising from the base of the stems. The flower-heads contain several black spikes closely packed, and surrounded by a long brown bract; May to July. Widespread in fens, marshes, and springs, particularly near the sea. **White Beak-sedge** *Rhynchospora alba* has leafier stems and very pale brown, almost white flowers.

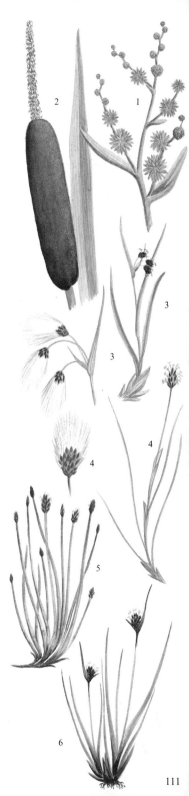

111

Bulrush *Scirpus lacustris* (1) is a very tall, rather rush-like perennial, with round, leaf-less stems ending in a branching head of reddish brown spikelets, with a short bract continuing the stem; June to July. A wide-spread, often common plant· of still and slow fresh water, often in quite deep water. **Glaucous Bulrush** *S. tabernaemontani* is bluish green and has 2 (not 3) stigmas in each flower. Commoner· near the sea.

Sea Club-rush *Scirpus maritimus* (2) is a tall perennial with triangular stems which bear long, narrow leaves. The stems end in tight clusters of dark brown spikes, set in 2 or more long, green bracts; July to August. Widespread in brackish and salty water. **Wood Club-rush** *S. sylvaticus* has spreading, loose clusters of smaller, greenish brown spikelets. Scattered in marshy woods, be-coming rare in Scotland.

SEDGES   True sedges *Carex* are all peren-nials (see p.109).
Greater Pond Sedge *Carex riparia* (3) is tall, tufted, and massive-looking. It has very long, rough-edged leaves, which bend in the middle. The upper spikes are male with black glumes and yellow anhers, the lower female with purple–brown glumes; May to June. Common in lowland marshes, be-coming scarce in Scotland. **Lesser Pond Sedge** *C. acutiformis* has less sharply pointed male glumes.

Greater Tussock Sedge *Carex paniculata* (4) is recognisable from afar because of the enormous tussocks it produces, sometimes well over 1 m high. It has very long, narrow leaves, and a flower-head in which all the spikes are similar and a dull orange–brown; June to July. Widespread in lowland areas in wet places and in alder woods.

Bottle Sedge *Carex rostrata* (5) has short to tall stems, depending on the fertility of its habitat. It is distinguished by the long, cylindrical, conspicuously yellowish female spikes, contrasting with the rather puny male ones; June to July. Common in Scot-land and N England in marshes and swampy lake-edges; scarce in the S.

False Fox Sedge *Carex otrubae* (6) forms short to tall, dense tufts, with stiff, upright leaves, up to 10 mm wide. The flower-head has several short, upright spikes, all appar-ently similar, though the top ones are in fact male; June to August. A common low-land sedge in damp soils and by streams and ditches.

Pendulous Sedge *Carex pendula* (1) is a tall or very tall, tufted sedge with long, flat, yellowish leaves up to 20 mm across. The topmost, brown, spike comprises of male flowers, the rest are long (up to 150 mm), slender and drooping, and greenish brown until mature; May to June. Frequent in damp woods, mainly in S and E England.

Wood Sedge *Carex sylvatica* (2) makes short, pale green tufts, with leafy stems bearing a single, very slender male spike at the top, and 4–5 narrow female spikes, which dangle on long, weak stalks; May to July. Common in damp woodland soils. **Ribbed Sedge** *C. binervis* has fatter female spikes on shorter stalks. Heaths, moors, and upland grassland, except in central and E England. **Hairy Sedge** *C. hirta* has hairy, upright spikes, and hairy leaves. Lowland grasslands.

Glaucous Sedge *Carex flacca* (3) is very variable in size but forms sparse tufts of greyish leaves, particularly on the underside. It typically has 2 narrow, brown male spikes, and a few more cylindrical female spikes, with a conspicuous contrast between the greenish fruit and the very dark scales; May to July. Very common in both wet and dry grassland. **Carnation-grass** *C. panicea* has a single male spike and few-flowered female spikes with fatter fruits.

Common Sedge *Carex nigra* (4) is a short, creeping sedge forming tufts of long, narrow leaves. It has 1–2 slender male spikes, and is best recognised by the very dark scales on the upright, short-stalked female spikes; May to July. Unlike most sedges with distinct male and female spikes, the female flowers have 2 (not 3) stigmas. Very common in marshes, fens and wet grassland.

Spring Sedge *Carex caryophyllea* (5) forms low to short, dark green tufts of flat, often curled leaves. The shape of the flower-head is distinctive, with 2 or 3 small female spikes clustered around the base of the solitary male spike; April to May. The scales of the female spikes are brown. Common in dry grassland, particularly on chalk, and in damper places in hills. **Pill Sedge** *C. pilulifera* forms denser, more yellowish green tufts. Heaths and acid grassland.

Sand Sedge *Carex arenaria* (6) is an obvious creeping sedge, its underground stems growing straight across sand-dunes, and producing well-spaced aerial shoots. The spikes are all similar and a clear pale brown; June to July. Common and sometimes the only plant on sand-dunes, particularly bare sand. **Brown Sedge** *C. disticha* is a very similar plant of inland marshes.

113

GRASSES. For terms used in describing grasses, see p.109.

Common Reed *Phragmites australis* (1) is a very tall perennial, sometimes over 2 m high, with long, thick, creeping underground stems. The stems bear long, flat leaves, up to 30 mm wide, held stiffly away; they have no ligule. The flowers are enclosed in long, brownish white hairs, and are in large, branching clusters; August and September. The stems remain through the winter, often in huge stands, in fens and shallow water.

Purple Moor-grass *Molinia caerulea* (2) is a tussock-forming perennial which has flat, dull green leaves with no ligule, only a fringe of hairs. The flowers are in long, branching spikes, each branch bearing several green, brown, or purplish spikelets, 5–9 mm long; July to September. Very common on wet, peaty soils.

Heath Grass *Danthonia decumbens* (3) is a short, tufted perennial with rather short, stiff leaves and no ligule, only a fringe of hairs. It has, however, a quite different flower-spike from Purple Moor-grass (above), with fewer, more rounded spikelets, 8–10 mm, whose glumes are at least as long as the rest of the spikelet, so seeming to embrace it; July to August. Common in poor grassland.

Meadow Fescue *Festuca pratensis* (4) forms tall, spreading tufts, which retain the brown remains of dead leaf-sheaths at the base. It has flat, rather narrow leaves, and a widely-branching flower-head. The branches are in pairs, one with several and one with only a single spikelet 10–15 mm long; June to July. A common grassland plant, usually on moist soils, becoming scarce in Scotland. **Tall Fescue** *F. arundinacea* is taller; each of the pairs of branches has several spikelets.

Red Fescue *Festuca rubra* (5) is a very distinctive grass, forming dense clumps of extremely fine, almost bristle-like leaves, which are often dark green. The leaf-sheaths are closed around the stem. The usually reddish spikelets are 8–12 mm long and have short awns; May to July. Extremely common in poor grassland. **Sheep's Fescue** *F. ovina* is very similar but has open leaf-sheaths.

Ryegrass *Lolium perenne* (6) is the most widely planted of all agricultural grasses. It has red or purplish stem-bases, small lobes which project almost around the stem at the base of each leaf, and flower-heads which are unbranched and have 2 alternating rows of unstalked spikelets, 10–15 mm; May to August. Very common. **Italian Ryegrass** *L. multiflorum* has its youngest leaf rolled, not folded. Also cultivated.

Reed Sweet-grass *Glyceria maxima* (1) is a very large grass which has creeping underground stems and forms large patches. Its leaves are very long and (for a grass) wide, up to 20 mm. It has very large, branching flower-heads, with narrowly oval, long-stalked spikelets, 5–10 mm long; June to August. Common in water and very wet places in S, E, and central England, very scarce elsewhere.

Floating Sweet-grass *Glyceria fluitans* (2) is one of the few truly aquatic grasses, and is most often seen with its rather short, flat leaves actually floating on the water. It has a little-branched flower-head, with very narrow spikelets, 20–30 mm long; May to August. Very common in shallow water or in grassland that is flooded in winter.

Annual Meadow-grass *Poa annua* (3) is a short annual or sometimes perennial grass forming small, rather yellowish green tufts. The leaves may be flat or folded, are blunt-tipped, and have a long ligule. It has small, spreading, branched flower-heads, with the small, oval spikelets around the outside, not near the central stalk; all the year. Extremely common in disturbed ground and open grassland.

Smooth Meadow-grass *Poa pratensis* (4) is a creeping perennial with long, white stolons, above or below ground. It has smooth stems bearing flat, blunt-tipped leaves with short ligules. The flower-head has spreading whorls of branches bearing flattened oval spikelets, 4–6 mm long; May to July. Very common. **Rough Meadow-grass** *P. trivialis* has rough stems, a long, pointed ligule and smaller spikelets (2–4 mm).

Cocksfoot *Dactylis glomerata* (5) can be identified when not in flower by its flattened shoots and when flowering by its very distinctive, brush-like flower-heads. These are branched, but the spikelets are clustered into 1-sided bunches, and the pointed glumes give them a bristly appearance; May to September. It forms large tussocks in grassland and open woods and is one of the commonest lowland grasses.

Crested Dogstail *Cynosurus cristatus* (6) is a short, tufted, stiff-stemmed perennial, with narrow, long-pointed leaves and long, parallel-sided, pipe-cleaner-like flower-heads. The spikelets are 3–4 mm long and are arranged along a central axis; some of them are sterile; June to August. Very common in poor grassland, and often spotted because of the persistent dead flower-heads.

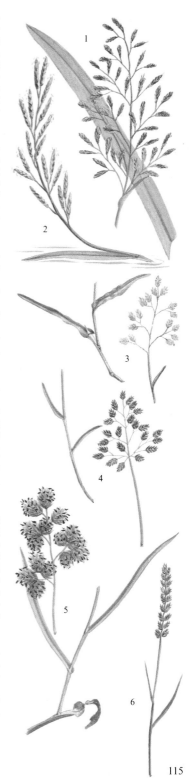

115

Upright Brome *Bromus erectus* (1) is a tall, tufted grass, whose long, narrow leaves, 2–3 mm wide, have short, rough-edged ligules. It has very narrow spikelets, up to 40 mm long, with prominent awns or bristles, in an upright, shortly branched spike; June to July. It is one of the most characteristic species of dry chalk and limestone grassland in S and E England, but it is very scarce elsewhere.

Hairy Brome *Bromus ramosus* (2) is a very tall, tufted perennial which has dark green, hairy leaves up to 15 mm wide and a much-branched, nodding spike of long, narrow spikelets, each with conspicuous awns; July to August. Common in woods and hedges, becoming scarce in the far N and W. **Giant Fescue** *Festuca gigantea* is hairless and usually grows on damper soils, though sometimes together with Hairy Brome.

Barren Brome *Bromus sterilis* (3) is a tall, downy annual with flat, often purple-tinged leaves which have a rather floppy feel. The flower-head bears numerous drooping branches, each of which bears a single, long-awned spikelet; May to August. Very common in disturbed ground in lowland areas, though rare N of Edinburgh.

Soft Brome *Bromus hordeaceus* (4) is a short or tall, very hairy annual, with flat, downy, greyish, rather floppy leaves, about 5 mm wide. It has a rather tight cluster of rounded, awned spikelets, 15–20 mm long, May to July. A very variable plant, which may be tall and luxuriant in meadows, and extremely dwarfed in dry, bare places.

Slender False Brome *Brachypodium sylvaticum* (5) is a tall, tufted perennial distinguished from the true brome grasses (above) by its unstalked spikelets, which are in 2 alternating rows along the spike; July to August. It has flat, soft, yellowish green leaves, and is common in woods and hedges. **Tor Grass** *B. pinnatum* has stiff green leaves and much shorter awns on the spikelets. Chalk and limestone grassland in S and E England.

Wood Melick *Melica uniflora* (6) is a patch-forming perennial that spreads through creeping underground stems and has rather short, flat leaves. The flowers are in a very sparse cluster, each of the purplish brown spikelets containing a single fertile floret; May to June. Common in woods, except in N Scotland.

Quaking Grass *Briza media* (1) is quite un-mistakable when in flower with its dangling spikelets, shaped like lanterns, on long, curved stalks; June to August. It forms loose tufts of flat, pale green leaves and is common in poor grassland, both dry and wet, except in the far NW. The only species with which it might be confused is Wood Melick (p.116), which has only one floret in each spikelet.

Couch Grass *Agropyron repens* (2) is a creep-ing perennial, whose thick white under-ground rhizomes are familiar to every gar-dener. It has flat leaves and a long, upright spike of unstalked spikelets, each 10–15 mm long; June to September. A common and pestilential weed. **Sea Couch** *A. pungens* is bluish grey and has strongly ridged leaves. On sand-dunes and in salt-marshes; not in Scotland.

Wall Barley *Hordeum murinum* (3) resembles a small barley plant, as it is rarely more than 50 cm high. It is an annual with short, almost triangular leaves and a dense flower-spike, composed of groups of 1-flowered spikelets, each with a long awn, 20–25 mm long; May to August. Common on disturbed ground, particularly by roads. **Sea Barley** *H. marinum* is very similar and grows in grassy places by the sea.

Meadow Barley *Hordeum secalinum* (4) is a usually short perennial with narrow leaves and much shorter awns (and so a less 'barley-like' flower-spike) than Wall Barley (above). It has a somewhat flattened spike; June to July. A widespread grass of lowland meadows SE of a line from the Humber to the Severn, but rare to the N and W. **Wood Barley** *Hordelymus europaeus* is a scarce plant of chalk and limestone woods.

Wild Oat *Avena fatua* (5) is a tall annual with broad, flat leaves and drooping spike-lets, enclosed by long glumes and with con-spicuous, projecting awns. It is a common and damaging weed of cereal crops and can be distinguished from the cultivated Oat (p.257) by the smaller spikelets and the red-dish brown hairs on the florets.

False Oat *Arrhenatherum elatius* (6) is a tall perennial with long, broad, flat, rough leaves. It has a loose, nodding flower-head, with 2-flowered spikelets, open when in flower. One of the florets in each spikelet has a long, twisted awn; May to August. A very common plant, typically found in grass-lands that have been disturbed in the past, and on roadsides.

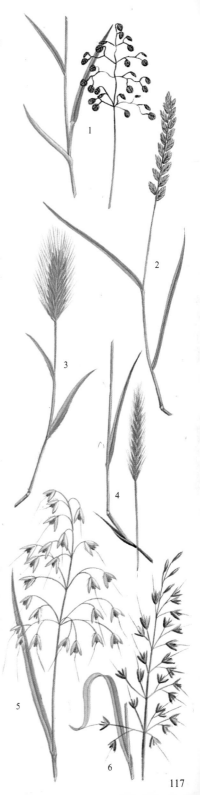

Yorkshire Fog *Holcus lanatus* (1) is a tufted perennial whose stems and leaves are uniformly clothed with soft hairs; the stems are striped at the base. The flower-head may be very tight or widely spreading, and pale green or pinkish, but it is always much-branched with small, papery, 2-flowered spikelets; May to August. Very common in grassland and open woods. **Creeping Soft-grass** *H. mollis* is greener and its stems are hairy only at the joints.

Tufted Hair-grass *Deschampsia caespitosa* (2) forms large tussocks of stiff, rough-edged, dark green leaves. It produces very tall flowering stems with spreading clusters of small (5 mm), 2-flowered spikelets, usually silvery green, sometimes purplish; June to July. Extremely common in damp grassland and abandoned fields, often among rushes, and preferring mineral soils to peat.

Wavy Hair-grass *Deschampsia flexuosa* (3) is an elegant, tufted perennial with wiry, rolled leaves, less than 1 mm wide but sometimes long. The flowering spike is similar to that of Tufted Hair-grass (above), but the spikelets have 2 awns and are silvery brown; June to July. A widespread species of heaths, moors and open woods, on very acid, usually dry, sandy soils; common on spoil heaps.

Marram Grass *Ammophila arenaria* (4) is a large perennial which forms extensive patches by means of creeping underground stems. It has very long, grey–green leaves, apparently narrow but actually rolled, 5–6 mm wide and distinctly ribbed when unrolled. The flowers are in a dense spike, composed of many 1-flowered, long, narrow spikelets; July to August. The commonest large sand-dune grass, often planted to stabilise sand.

Creeping Bent *Agrostis stolonifera* (5) is a creeping perennial whose stems are at first flat along the ground and then bend upwards. The stems bear flat, pointed leaves which have long ligules. The spikelets are 1-flowered and only 3 mm long, in an open, branching cluster, which becomes tight and spike-like in fruit; July to August. Very common in grassland and on waste ground, preferring damp soils.

Common Bent *Agrostis tenuis* (6) has short tufts and rarely creeping stems. Its leaves are usually narrower than Creeping Bent and it has a short, rounded ligule. The flower-head is much more open and often purple-tinged. Very common in mildly acid, usually dry grasslands. **Brown Bent** *A. canina* has a long, pointed ligule and tufts of leaves from each stem-joint.

Meadow Foxtail *Alopecurus pratensis* (1) is a tall perennial with long, flat, pointed leaves which have short, rounded ligules. The spikelets are in a very dense spike, which is green, but with conspicuous purple anthers; April to June. Each spikelet has a single floret and a short awn. Foxtail is a very common grass of lowland meadows on better soils.

Marsh Foxtail *Alopecurus geniculatus* (2) is a creeping perennial with weak-kneed stems, eventually upright, but rooting at the joints. It has flat, rather greyish leaves, with long, blunt ligules, and a flower-spike very like that of Meadow Foxtail (above), but shorter, up to 60 mm long; June to August. It is very common in wet grasslands and on muddy water-edges.

Timothy *Phleum pratense* (3) is a very tall, tufted perennial, whose broad, flat, pointed leaves have long ligules. It has a long, cylindrical flower-spike, up to 30 cm long in cultivated forms, that is less tapering than that of Meadow Foxtail (above) and usually a distinctive greyish green; June to July. Common in damp grasslands and widely cultivated, often as very vigorous cultivars.

Reed Grass *Phalaris arundinacea* (4) is a very tall perennial with creeping underground stems, whose fat stems bear flat leaves, 10-15 mm wide. The flower-head is branched, with the branches bearing dense clusters of 1-flowered spikelets, but themselves well-spaced, giving the appearance of a large Cocksfoot (p.115); June to July. Common in wet grassland, marshes, and damp woods.

Sweet Vernal Grass *Anthoxanthum odoratum* (5) is a short, aromatic, tufted perennial whose flat leaves are broad for their length and have long ligules. The flowers are in short spikes of short-stalked, 3-flowered spikelets (only one of which actually produces a seed); April to June. The seeds are conspicuous – very dark brown and attached to a twisted awn. A very common grass of grassland, heaths, and open woods, on a range of soils.

Common Cord-grass *Spartina anglica* (6) is a tufted perennial with stiff, greyish leaves and a characteristic flower-head, comprising several stiffly upright spikes, each composed of many overlapping, 1-flowered spikelets; July to October. Although now much the commonest grass on tidal mud-flats, this species evolved in the late 19th century by hybridisation of an introduced American and a native species.

Lower Plants and Fungi

life cycle of lower plants

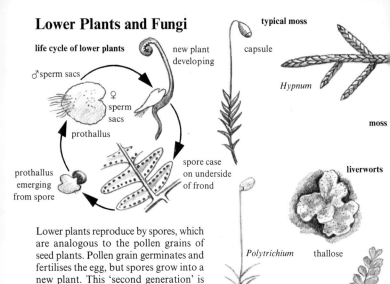

typical moss

capsule

Hypnum

moss

♂ sperm sacs

♀ sperm sacs

prothallus

new plant developing

prothallus emerging from spore

spore case on underside of frond

Polytrichium

liverworts

thallose

leafy type

Lower plants reproduce by spores, which are analogous to the pollen grains of seed plants. Pollen grain germinates and fertilises the egg, but spores grow into a new plant. This 'second generation' is an insignificant green disc in ferns, but in the mosses is what we call the moss, and the spore-bearing generation is the capsule, growing parasitically on the 'second generation'.

Ferns are mainly large plants, with a well-developed vascular system in their stems for transporting water, and feathery leaves. But some are smaller and some of their relatives look very different. Horsetails are much simpler in struc-

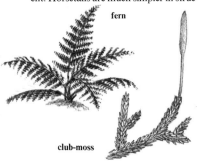

fern

club-moss

ture and are the last survivors of some of the first land-plants, which first appeared nearly 400 million years ago. Club-mosses are related to ferns, though they look mossy.

Mosses and liverworts tend to grow in wetter places than ferns and have no vascular system. Some grow in dry places by the simple expedient of drying out, but cannot control their water loss. Mosses can be acrocarpous, with short, upright shoots, and pleurocarpous, which are largely creeping. Similarly there are leafy liverworts, with 2 or 3 ranks of flimsy green leaves on a creeping

stem, while thallose liverworts consist of a flat green plate, the 'thallus'.

Fungi are a separate kingdom, distinguished by the way they obtain food, by digesting organic matter, whether dead or alive, when they become parasites. They reproduce by spores but have no second generation. Some fungi form an extraordinary association with tiny green algae, such as those which coat tree-trunks, and produce a lichen.

Lichens can withstand very severe conditions (drought, cold, heat, etc.). Many of them look like thallose liverworts, but they are much tougher and rarely green – these are the frondose types. Others are fruticose, and they and the crustose types, which form crusts on rocks and tree-trunks, are hard to mistake for anything else.

lichens

Peltigera

Parmelia

Xanthoria

mushroom

HORSETAILS *Equisetum* are very primitive plants which consist of long, green stems, sometimes unbranched but in most species with whorls of thin, green branches. The spores are borne in brown, cone-like structures, which may either be on the ends of the green stems or on separate brown ones.

Common Horsetail *E. arvense* (1) produces brown fertile stems in early spring and later develops short to tall sterile stems, deeply grooved and bearing whorls of simple teeth. A common and persistent weed; also in grassy places. **Giant Horsetail** *E. telmateia* also has separate fertile stems; it is much larger and grows in damp places.

Water Horsetail *Equisetum fluviatile* (2) has very long, thin stems with uneven whorls of very short branches, and many fine grooves. Some of the stems taper to a point, others end in a brown 'cone'. Common in shallow water. **Marsh Horsetail** *E. palustre* is smaller and has a few deep grooves on the stem. Marshes and wet meadows.

FERNS Polypodiaceae

Bracken *Pteridium aquilinum* is the commonest British fern. It has a dark, creeping underground stem from which arise tall to very tall fronds, upright at first and then branching and flatter. Each frond is 3 times pinnate and the spores are borne in a continuous rim around the leaf segments. Very common in woods and grassland, preferring dry, acid soils, and sometimes dominating huge areas.

Royal Fern *Osmunda regalis* is a large, mound-forming fern, producing each year very long twice-pinnate leaves, the innermost of which have their topmost leaf-lobes modified to produce spores. It is very long-lived and eventually comes to sit on a mound as much as 1 m high. Widespread in the W on damp peaty soils; rare in the E.

Rusty-back Fern *Ceterach officinarum* is a low, tufted fern with short, simply pinnate leaves which are green on the top but brown underneath because of the presence of many overlapping scales. Cracks in walls and limestone rocks, mainly in the S and W.

Hard Fern *Blechnum spicant* is a distinctive fern with separate fertile and sterile leaves. The ordinary leaves form tight clumps and are bright green and once-pinnate, with the longest leaflets near the middle of the leaf. The fertile leaves arise near the centre and have very narrow leaflets with spores produced on the underside; they too are green. Common in grassy and heathy places and woods on acid soils, and on rocks.

121

Maidenhair Spleenwort *Asplenium tricho-manes* (1) is a low, tufted fern whose leaves have a long black midrib and many (up to 40) pairs of small oval leaflets. The spores are in groups on the veins on the underside of the leaves. Very common on walls and rocks. **Green Spleenwort** *A. viride* has a green midrib. Upland areas.

Wall-rue *Asplenium ruta-muraria* is very small but highly distinctive with its deep, dull green leaves, rather irregularly twice-pinnate or trefoil. The leaflets are usually oval, but widest near the top, and bear small clusters of spores underneath. Very common on walls and rocks.

Hart's-tongue Fern *Phyllitis scolopendrium* is a tufted fern with unlobed, yellowish green leaves on short stalks. The underside of each leaf is striped brown with pairs of lines of spore-sacs. Common in the W and S, becoming more scarce in the drier E and the far N; on rocks and banks often in deep shade, in woods, caves, and wells.

Male Fern *Dryopteris filix-mas* (2) forms large clumps of very long, twice-pinnate leaves, with the longest primary branches in the middle of the leaf, giving the whole leaf a more or less oval shape. The spores are in rounded clusters under the leaves. Common in woods, hedges, and on rocks. **Lady Fern** *Athyrium filix-foemina* is more delicate in appearance and has curved spore-clusters. It prefers acid soils.

Broad Buckler Fern *Dryopteris dilatata* (3) differs from Male Fern (above) in having 3-times pinnate leaves which have long stalks and only bear leaf-branches for about half the length of the whole leaf. The stalk is covered with brown scales which have a dark area in the middle. Common in damp woods and heaths, and in shady cracks in rocks. **Narrow Buckler Fern** *D. carthusiana* is usually paler green and the scales on the leaf-stalk are uniformly pale brown. It forms a less spreading clump. Mainly in wet woods.

Polypody *Polypodium vulgare* is a short, tufted fern with rather leathery, pinnately lobed leaves on long stalks. The spores are in large yellowish clusters beneath the leaves. Very common in the wetter parts of Britain on rocks, walls, and, where it is wet enough, on trees; scarce in the E. It is a larger plant than Rusty-back Fern (p.121) and does not have the rusty undersides to the leaves.

CLUBMOSSES are actually relatives of the ferns, not the mosses (see p.120), and can be distinguished from mosses when fertile by the spores being in clusters at the leaf-bases (in Fir Clubmoss) or in horsetail-like 'cones' (the other species).

Fir Clubmoss *Lycopodium selago* (1) is low, bristly and upright, like a tiny fir tree, with the orange spore clusters at the base of the upper leaves. **Stagshorn Clubmoss** *L. clavatum* (2) has long, creeping stems and long-stalked cones. Both are common in mountain grassland, scarce elsewhere.

MOSSES

Dicranella heteromalla (3) is very small, only a few cm high, and has curved, narrow leaves which all point to one side. The capsule turns orange and bends sideways when ripe, on a long (10–15 mm) yellowish stalk, Common on bare, acid soil, rocks, and in woods. *Campylopus pyriformis* is similar but the capsule is pale brown. *Dicranum scoparium* is much larger, up to 10 cm high.

Sphagnum palustre is one of a large and extremely important genus responsible for forming much of the peat of W Britain. Mats of *Sphagnum* are recognised by their fleshy appearance and their pointed leafy branches. Their unique leaf structure allows them to retain water and so create the water-logged conditions in which they thrive. The individual species can only be identified by experienced naturalists.

Polytrichum commune (4) is a large, stiffly upright moss, sometimes as much as 20–30 cm tall. The stiff, pointed, and minutely toothed leaves are held almost horizontally and it produces abundant square capsules, each with a pointed lid. It is very common in wet moors and open woods. *P. formosum* is smaller and has a 5- or 6-sided capsule; it is common in woods. *P. juniperinum* has un-toothed leaves with long, brown tips.

Mnium hornum (5) is a common woodland moss, often growing on thick leaf-mould. It has flat, deep dull green leaves (except for the new shoots which are paler) on very short, upright stems. Male plants have a rosette of leaves at the top; females eventually produce a thin, cylindrical capsule on a long, thin, bent-topped stalk. *M. punctatum* is larger, with pale green, oval leaves. *M. undulatum* has long, narrow, wavy leaves.

Funaria hygrometrica has stems less than 1 cm tall, bearing a few pointed, oval leaves. The capsule is a twisted pear shape, on a very long, curved stalk. It grows on bonfire sites and other bare areas.

123

Thuidium tamariscinum (1) is a very characteristic, bright pale green, creeping moss whose stems are 3 times pinnate, giving the whole frond a feathery appearance. It is common in woods and hedges, sometimes in deep shade. *Hylocomium splendens* is another much branched moss, but it is only twice pinnate and has reddish stems.

Hypnum cupressiforme (2) is probably the commonest British moss. It forms dense mats on the ground, on tree-trunks, on walls, and in a variety of other habitats. The shoots are simply branched and are densely clothed with curved, overlapping, long-pointed leaves, but this is a very variable species and its varieties and closely related species are hard to separate. *Pleurozium schreberi* is a superficially similar plant with blunt leaves and red stems; it grows on heaths and in acid woods.

Eurhynchium praelongum (3) is a very common woodland moss, often found growing on tree-trunks. It forms spreading mats which are best recognised by the much larger leaves on the main stems than those on the branches. *Rhytidiadelphus squarrosus* is a grassland plant with equal leaves which are curved and held away from the stem, giving a bristly appearance, and *R. triquetrus* is similar, with straight leaves and stiff, red stems.

Brachythecium rutabulum is a large, creeping, raggedly branched moss with spreading, glossy green leaves and abundant, long-stalked capsules. It is a very common moss in a wide range of habitats, including woods, grasslands and bare soil, usually in fertile conditions.

Pseudoscleropodium purum is one of the most easily recognised mosses with its rather fat, pale green shoots, tightly clothed in broad, pointed leaves. The shoots are very typically blunt-tipped. It is a common moss of grassy and wooded habitats and it very rarely produces capsules.

Fontinalis antipyretica (4) is a very familiar sight, forming long, very dark green, trailing stems in flowing water, and often completely clothing stones and tree-roots. Out of the water it is best recognised by the folded leaves. *Cinclidotus fontinaloides* is a similar but much smaller moss that grows in the same habitat, but its stems are rarely more than 20 cm long, whereas those of *Fontinalis* are typically 2–3 times that length.

Ceratodon purpureus is an extremely common but very variable moss. It is the commonest small (2–3 cm), upright (acrocarpous) moss and grows in a wide range of habitats, including soil, walls and fallen trees. It has small, narrow oval leaves usually held almost horizontally, giving it a 'spiky' look, and in spring bears abundant, long-stalked, purple capsules. It is common in towns.

Bryum argenteum (1) is one of the few species in this large and difficult genus that is readily identified. It has characteristic silvery grey leaves tightly pressed against the stem to resemble an upside-down catkin, usually only about 10 mm high. Its capsules are small and inverted on stalks about as long as the shoot. It is a common urban moss, on pavements and little-used tarmac and concrete. *B. caespiticium* is one of the commonest green *Bryum*s, particularly on walls and tree-trunks.

Grimmia apocarpa (2) is a common wall-top moss, distinguished from the *Bryum*s (above) by the long, greyish points to the leaves, which are particularly conspicuous when dry. It varies in colour from green through a very dark grey–green to almost brown, and the capsule is distinctive in being almost unstalked. *G. pulvinata* has longer points, stalked capsules, and forms rounded, grey cushions.

Tortula muralis forms very short cushions only a few millimetres high on rocks and walls. It is distinguished from other mosses in this habitat (above) by its rounded-tipped leaves which end in a long, thin hair, and by its long cylindrical capsule which is quite vertical. When dry the clumps are greyish but turn a clear, pale green when moist.

Atrichum undulatum is a large moss, up to 5 cm tall, and somewhat resembling a *Polytrichum* (see p.123), but quite different from them in its pale green, wavy-edged leaves, like those of *Mnium undulatum* (p.123). Its long-stalked, curved capsule has a long, pointed beak. It is common on soil in woods.

Leucobryum glaucum is easily the most distinctive British moss. It forms large cushions which when dry are almost white, and even when moist are normally a very pale bluish green. Each shoot has long, curved leaves, but the dead lower parts remain in the cushion and contribute to the white appearance. It is found in shady, acid woods and on wet moors.

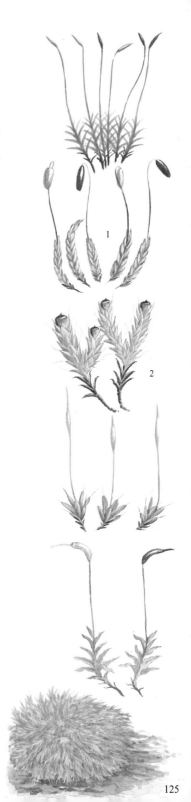

LIVERWORTS

Lophocolea heterophylla (1) consists of clumps of short, branching stems a few centimetres long, bearing 2 ranks of pale green leaves, some rounded and some notched. In spring it produces dark capsules on pale, fleshy stalks. It is very common on dead wood and tree-trunks in woods, typically among mosses. *L. cuspidata* and *L. bidentata* both have all their leaves deeply notched; the latter grows on the ground.

Marchantia polymorpha is one of the most familiar 'thallose' liverworts, for it commonly grows in greenhouses and nurseries as well as on overhanging river-banks. The dark green thallus is spotted with round cups and the spores are produced on the underside of strange, umbrella-like structures, which may have 9 stubby rays (female plants) or a flat disc (male plants).

Lunularia cruciata is a very pale green liverwort that is most often encountered in gardens and in greenhouses, sometimes achieving weed status. It may be distinguished from *Marchantia* (above), the other common thallose liverwort in gardens, by its paler colour, the crescent-shaped (not circular) cups on its surface, and the fact that it rarely bears spore-producing structures.

Pellia epiphylla is a rather large liverwort that is common on damp banks and by streams, and is best distinguished from the otherwise similar thallose liverworts on this page by its lack of their distinctive features – the round cups and 'umbrellas' of *Marchantia*, the crescents of *Lunularia* (above) or the spotted surface of *Conocephalum* (below). It bears round, shiny, black capsules on short, translucent stalks in late spring.

Conocephalum conicum is one of the largest of the common thallose liverworts, producing branching ribbons up to 20 cm long, and forming extensive patches on wet rocks and walls. It is at once identified by the spotted appearance of the thallus, which on close examination is seen to be caused by the whole surface of the thallus being divided into hexagons with a small pore in the centre of each.

Plagiochila asplenioides is a common and luxuriant leafy liverwort of moist, shady places. The stems bear 2 rows of rounded, untoothed leaves which overlap to such an extent that at a glance they seem continuous. It forms large, dense, rather upward-growing clumps.

126

LICHENS

These are curious compound organisms, composed of a fungus which can only survive in the association, and an alga which is usually a common form, such as those that form green films on tree-trunks or damp soil. The structure which results from the association of these two partners is quite unlike anything that either can produce on its own. There are 3 basic types.

Some lichens have more or less tubular, branching and upright forms, and perhaps the best known of these are the *Cladonias*. *Cladonia furcata* (1) is a very common greeny brown lichen of damp, peaty soils, whose forked stems are sometimes tipped with red spore-bearing structures. Other common species are *C. coniocraea*, a bluish green species of tree-trunks with unbranched stalks, and *C. floerkeana* with conspicuous large red spore-bodies. *Ramalina farinacea* (2) also has branching stalks, but they tend to be flattened and the whole is a dull yellowish green or brown. It is one of the commonest of tree-trunk lichens. *Usnea subfloridana* (3) is another branching lichen of tree-bark, but is extremely sensitive to air pollution and is only to be found in the W and N, where it is common. It forms dangling, tangled masses of much-branched stems, which are a characteristic greyish green. *Evernia prunastri* (4) resembles a grey –green *Ramalina* and is also found on trees, though additionally on fence posts and on rocks in wet areas. Its flattened, forking branches are very distinctive.

Many lichens have a quite different structure, and they form more or less rounded, flattened, leaf-like plant-bodies, somewhat resembling a thallose liverwort, but never the fleshy, pure green of those. *Hypogymnia physodes* (5) is one of the commonest lichens on tree-bark and it is relatively resistant to air pollution. Its fronds are lobed and silvery grey. *Peltigera canina* (6) is a ground-living lichen, often found amongst grass on wet lawns. It is a large lichen with dull grey – brown fronds which have root-like growths underneath. *Xanthoria parietina* (7) is an instantly recognisable species from its colour alone. It is very common on rocks, particularly by the sea, and on roofs.

The third sort of lichen is the type that forms amorphous crusts on rocks and trees, and perhaps the most familiar of these is *Lecanora conizaeioides* (8). It forms grey–green patches on tree-trunks, walls and other surfaces; one of the few lichens resistant to air pollution, it is common in towns.

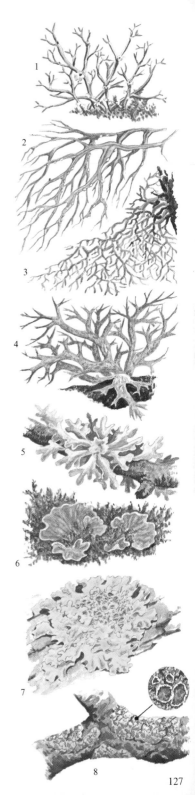

127

Orange Peel *Peziza aurantia* is perhaps the commonest of a very large group of fungi called the cup fungi, from the shape of the fruiting bodies. It is bright scarlet–orange inside the cup, paler and felted outside, and is common on bare ground from September to January. Two other common woodland species are *P. haemisphaerica*, which has a small cup, to 2 cm across, brown outside and paler within, and *P. succosa*, slightly larger, brown outside and purplish brown within.

Morel *Morchella esculenta* is a large species, up to 12 cm high, with a somewhat oblong, conical cap, joined to the stalk, and irregularly pitted all over like a honeycomb. It appears in spring (March to May) in woods and grassland on good soils, and is edible and good. Several other morels also occur, but *M. esculenta* is the most frequently encountered, particularly in SE England.

Candle Snuff *Xylaria hypoxylon* (1) consists of many tough, often forking, irregularly shaped fingers growing out of dead wood. They are black, though white-tipped when young, the white being the spores. Extremely common on dead wood throughout the year. The rather similar **Earth Tongues**, such as the black *Geoglossum hirsutum* or the dark olive-green *G. viride*, grow in soil in grassland.

Cramp Balls *Daldinia concentrica* form globular excrescences on the dead branches of broad-leaved trees. They are chocolate-brown when young but blacken with age; if cut across they reveal concentric rings (hence the Latin name). A very common species which can be found throughout the year.

Orange Spot *Nectria cinnabarina* (2) is seen as a rapidly spreading cluster of tiny orange, pinkish or red spots, a few millimetres across, on dead or dying wood of all kinds. It can be found throughout the year. Other common species include **Apple Canker** *N. galligena*, on apple trees, and *Diatrype disciformis*, on beech; both are brown.

Horn of Plenty *Craterellus cornucopioides* is a dark brown, funnel-shaped fungus, with no obvious distinction between stalk and cap. It can be up to 12 cm high and has a wavy edge to the cap. The underside is wrinkled and has no obvious gills. It can be found on the ground in beechwoods and under other deciduous trees from August to October.

Chanterelle *Cantherellus cibarius* is one of the finest edible fungi. It is very variable in size, up to 10 cm high and the same across, with a dish-shaped cap and gills running down on to the stalk. The whole fruiting-body is a pale apricot-yellow. It can be found in broad-leaved woods from July until the first frosts.

Razor Strop *Piptoporus betulinus* is much the commonest bracket fungus and grows almost exclusively on birch trunks. The brackets burst out from the bark and may live for several years. They are pale, often greyish brown and smooth on the top surface, and the underside is creamy white and has hundreds of tiny pores containing the spores. If the angle of the trunk is altered the pores grow so as to remain perfectly vertical.

Many-zoned Fungus *Trametes versicolor* is a bracket fungus usually seen flush with the surface of fallen branches of broad-leaved trees. The brackets are thin and may be circular or semi-circular. The upper surface is velvety and has concentric rings of almost any colour from olive-brown to black. It can be found throughout the year.

Giant Bracket *Grifola gigantea* forms a massive tuft of caps, often as much as 1 m across, arising from near the ground on the trunks of broad-leaved trees. Each cap is brown with a rather loose surface and a short, fat stalk, and the undersides are a mass of tiny, white pores. They can be found at almost any time of year except spring and early summer.

Beefsteak Fungus *Fistulina hepatica* is a curious bracket fungus resembling a large tongue more than a beefsteak. It is rough-surfaced and reddish brown above, and has minute pores on the underside. When cut it bleeds a dull red juice, and the cut flesh resembles meat (hence the name). A widespread species most often on oaks, from August to late autumn.

Oyster Fungus *Pleurotus ostreatus* is most often seen on beech trees, growing out of the living trunk. It has a blue–grey cap when young, but this becomes brown with age. The stalk is well to one side of the cap and the white gills run down on to it. An edible species that can be collected throughout the year.

129

Honey Fungus *Armillaria mellea* is an extremely common and very damaging parasite of living trees. It grows in tufts in woods of all types and spreads by thick black threads called rhizomorphs. Its cap is yellow-brown and slightly scaly when young, and the conspicuous ring on the stem is yellow or yellow-spotted. An edible fungus found in late summer and autumn.

St George's Mushroom *Tricholoma gambosum* (1) is a very large fungus, with a creamy-white cap up to 15 cm across. The gills are white, narrow and often wavy. It has a short, fat, often curved stalk. An edible fungus found in fertile pastures in spring. *Clitocybe* is a large genus of rather similar, but mainly woodland species, whose gills run down on to the stem; some are poisonous.

Wood Blewits *Tricholoma nudum* (2) is a medium-sized fungus with a rather flat, brownish violet cap. The narrow, crowded gills are pale purple. A common edible toadstool in late autumn in woods and grassy places. **Blewits** *T. saevum* has a paler cap and bluer stem and is rarely found in woods. *T. fulvum* has a pointed brown cap and yellow gills; it is found on heaths and in birchwoods.

Laccaria *Laccaria laccata* is a very common, variable woodland fungus with few, widely-spaced gills and a distinctive browny orange colour. The cap is small, usually less than 5 cm across, but the stalk is usually long. Found in summer and autumn till early winter. Small species of *Clitocybe* may be confused with it; they have a larger number of more crowded gills which run down on to the stalk. Some are poisonous.

Fairy-ring Champignon *Marasmius oreades* is the toadstool that causes fairy rings on lawns by its habit of growing in an expanding circle and encouraging grass growth behind the growing front. It has small, slightly pointed caps and few, widely-spaced gills. This genus and the related *Mycena* include the host of very delicate tiny species found on leaf-litter and small twigs in woods.

Slimy Beech Tuft *Oudemansiella mucida* (3) typically grows on standing dead trunks of Beech. It occurs in groups of shining, slimy, pure white caps on very flexible stalks, each with a ring, in autumn. The gills are few and well-spaced. *O. radicata* grows on the ground in beechwoods. It has a long stalk and slightly pointed, brownish cap.

Fly Agaric *Amanita muscaria* is one of the most easily recognised toadstools, with its large, brilliant orange–scarlet cap flecked with white scales, and its white stalk with a prominent ring. Old specimens lose the red tint. The gills are well spaced and rather fragile. Fly Agaric is common in birch- and pinewoods in late summer and autumn and contains a powerful hallucinogen which is **poisonous**.

Tawny Grisette *Amanita fulva* is a medium-sized species with a dull orange–brown cap. The cap is unmarked with a slightly wavy edge, and sits on a slender stalk which has no ring. It can be found throughout the summer and autumn in woods, particularly birch-woods. **Grisette** *A. vaginata* has a yellowish grey cap and is found in late summer and autumn in woods and among heather.

Destroying Angel *Amanita virosa* is fortunately instantly recognisable and uncommon, for it is **deadly poisonous**. It is an entirely white toadstool – cap, gills, and stalk. The stalk has both a ragged ring near the top and a membrane around the rather bulbous base. It can be found in broad-leaved woods in the autumn.

Death Cap *Amanita phalloides* is the best known of British **poisonous** fungi. It contains 3 separate poisons and is lethal, even in small quantities. It can be recognised by its yellow–olive cap which is faintly striped, and the conspicuous ring at the top of the stalk. It also has a membrane around the bulbous base of the stalk. In a good year it is quite frequent in late summer and autumn, mainly under Oak and Beech.

Panther Cap *Amanita pantherina* is a medium-sized, rather squat toadstool with a greeny brown cap flecked with white scales. When mature the cap is almost flat, and the stalk has 2 or 3 rings around it. It is an uncommon species, found in autumn, particularly in beechwoods; **poisonous**.

Blusher *Amanita rubescens* has a yellowish brown or reddish brown cap with brownish white scales, but is best recognised by the purplish red colour developed wherever the flesh is damaged, e.g. by slugs. The stout, whitish stalk has a prominent ring and is often reddish near the base. A widespread woodland toadstool found in late summer and autumn.

poisonous

poisonous

poisonous

poisonous

Parasol Mushroom *Lepiota procera* is a large, long-stalked mushroom with a greyish beige cap. Both stalk and cap have darker scales. The stalk bears a conspicuous double ring. A widespread species of open woods and grassy places in late summer and autumn. The genus *Lepiota* is a large one; most species are smaller but similar in shape.

Field Mushroom *Agaricus campestris* (1) is perhaps the best-known edible mushroom. It is a rather stout, white species with pinkish brown gills, and a single ragged ring round the stalk, soon falling. A widespread grassland species in late summer and autumn. **Cultivated Mushroom** *A. bisporus* is the brownish white and much less tasty species, with browner gills, that is grown commercially.

Wood Mushroom *Agaricus silvicola* has an off-white cap, becoming yellow with age and turning yellow if bruised. The stalk has a hanging and rather frilly ring, and ends in a swollen base which does not turn yellow if bruised or cut. The gills are greyish brown and smell of aniseed. A common woodland species in late summer and autumn.

Horse Mushroom *Agaricus arvensis* is like Wood Mushroom in most respects but is usually larger and coarser. The stalk is straight and does not end in a bulbous swelling. It is primarily a grassland species and is rarely found in woods. It appears in late summer and autumn.

Yellow-staining Mushroom *Agaricus xanthodermus* resembles the previous 2 species in having flesh that turns yellow when cut or bruised, but in this species the whole fruit-body, including the base of the stalk, reacts in this way. It differs also in being inedible, often causing distressing symptoms. This species has purplish brown gills and a rather greyish cap.

Shaggy Ink-cap *Coprinus comatus* is an unmistakable fungus with a shaggy, white, cylindrical cap when young, which gradually opens from the base as the spores mature. The mature gills resolve into a soggy black soup. It is a very common grassland plant throughout summer and autumn, and if picked when young is good to eat. Several other species occur, often on dung.

Red-staining Mushroom *Inocybe patouillardii* is distinguished by its flesh turning pinkish red when bruised. The cap is a dull whitish buff where unbruised and is often split around the edge. It has a stout stalk with no ring. An infrequent species, particularly in beechwoods, in summer and autumn. **Deadly poisonous.** Many other species occur, mostly with grey– or orange–brown, pointed caps; many are poisonous.

poisonous

Common Cortinarius *Cortinarius cinnamomeus* is a small, dull ochre-coloured species with a rounded or sometimes pointed cap. The gills are yellow when young but turn brown. Very common in woods from late summer onwards. This is a very big genus with both small and large species; all are dull-coloured and have a very delicate veil between stalk and cap when young.

Cep *Boletus edulis* is the most edible of this very large and distinctive genus. All have stout, sometimes almost spherical stalks, and fat, fleshy caps. They do not have gills but a mass of tiny pores in which the spores are produced. Cep has a yellowish brown cap and a white-veined stalk. It is common in deciduous woods in autumn. *B. variegatus* is a slimy-capped, yellowish species of pinewoods; *B. scaber* has a striped stem and is found under Birches.

Sickener *Russula emetica* is a medium-sized fungus with a scarlet cap which breaks easily if picked. The gills are well-spaced and pure white, even when mature, and it has a long, white stalk. Common in coniferous woods in late summer and autumn. Harmless, but causes vomiting. Several other red *Russula*s are common, including *R. mairei* (beechwoods) and *R. lepida* (reddish stalk).

Yellow Russula *Russula ochroleuca* is very similar to Sickener but has a bright yellow cap. It is a common woodland species in autumn. Other yellow-capped *Russula*s include *R. solaris*, which is smaller, has yellowish gills and grows in beechwoods, and *R. claroflava*, with a lemon-yellow cap and pale yellow gills; it grows under Birch.

Dark Red Russula *R. atropurpurea* is similar in most respects to Sickener but its purplish red cap becomes extremely dark in the centre. The stalk is slightly brownish white. A common species of deciduous woods in late summer and autumn. Similar species include the smaller *R. nitida* (birchwoods) and *R. olivacea* (beechwoods); both have yellowish gills.

133

Saffron Milkcap *Lactarius deliciosus* is a small to medium, pinkish orange species with concentric markings on the hollow-centred cap. The cap and rather stout stalk are often green-stained, and when they are broken the juice turns orange. An edible species, often found under conifers in autumn. Many other *Lactarius* species occur, mostly dull reddish or orange–brown, for example *L. rufus*, under Pines.

Stinkhorn *Phallus impudicus* is quite unmistakable with its thick white stalk and black honeycombed cap, covered with sticky, foul-smelling jelly which contains the spores. The stalk rises from a jelly-filled, papery egg. A common species in deciduous woods in late summer and early autumn.

Giant Puffball *Lycoperdon giganteum* forms an enormous, off-white, smooth, leathery ball, typically up to 30 cm across, and occasionally more. The inside is spongy when young, but ripens to a chocolate-brown mass of spores, which puff away under pressure. Widespread in grassland in autumn. Several smaller species occur, including *L. perlatum*, which is yellowish brown and warty, and *L. pyriforme*, which grows on wood; both are pear-shaped.

Common Earth-star *Geastrum triplex* is a curiously shaped fungus which is spherical when young, after which the outer layer cracks into 4 or 5 arms to reveal the fruit-body within. A widespread and sometimes common plant of deciduous woods in autumn. Several other species occur, though they are less common; some have stalked fruit-bodies, giving the appearance of a manikin.

Common Earth-ball *Scleroderma aurantium* resembles a tough, dull orange puffball, with a thick networked skin. The spores are inside and when mature are black; they escape by the rupture of the skin. Earth-balls are common on the ground, often at the base of trees, and are found in birch-woods in autumn and early winter.

Witch's Butter *Tremella mesenterica* is one of the commonest and most conspicuous of a large group of gelatinous fungi. It consists of folded orange flanges bursting out from dead, usually coniferous wood. It occurs throughout the year. *Auricularia auricula* is larger and brown and is found on Elder trees.

Mammals

Mammals are the most advanced group of backboned animals, distinguished by their superior intelligence, by the hair on their bodies, and by feeding their young on milk secreted by the females in the mammary glands which give them their name. Eight major orders are represented in our wild fauna.

Insectivores Insectivora are the most primitive order of British and Irish mammals, characterised by their long snouts ing, in Britain and Ireland, 2 species of hare and the Rabbit.

Rodents Rodentia consist of a large group of mostly small vegetarian mammals, with teeth specially adapted for gnawing, including the squirrels, rats, mice, dormice, voles and the Coypu.

Cetaceans Cetacea (see p.255) comprise the whales, dolphins and porpoises, all adapted to spending their whole life in the water, their limbs being modified into flippers.

Carnivores Carnivora are predatory, flesh-eating mammals, represented in

shrew **Insectivores**

bat **Chiroptera**

mouse **Rodents**

Badger **Carnivores**

deer **Even-toed Ungulates**

Rabbit **Lagomorphs**

and numerous sharp teeth adapted to their preferred diet of invertebrates. Our 6 species are the Hedgehog, the Mole and 4 species of shrew.

Bats Chiroptera are highly specialised insectivorous mammals, with wing membranes which make them the only mammals capable of flying like birds. They are not blind, but use a form of echo-location to guide them at night. They roost in hollow trees, caves and buildings. There are a dozen species in Britain and Ireland.

Lagomorphs Lagomorpha are ground-living, herbivorous mammals, comprising Britain and Ireland by 10 species in 3 families.

Seals Pinnipedia are less exclusively adapted to marine life than the whales, since they come ashore, but are highly specialised nonetheless. There are 2 British and Irish natives.

Even-toed Ungulates Artiodactyla are one of the 2 orders of herbivorous hoofed mammals, whose wild representatives in Britain are the deer, and domesticated ones cattle, sheep and goats (pp 258–9). The horse (p.259) belongs to the **Odd-toed Ungulates** or Perissodactyla.

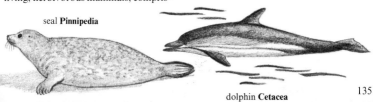

seal **Pinnipedia**

dolphin **Cetacea**

INSECTIVORES Insectivora

Hedgehog *Erinaceus europaeus.* Nowadays more familiar as a squashed corpse on the roads, the Hedgehog is a widespread and common animal, absent only from a few Scottish islands, that will come to feed at rural or suburban back doors if milk or other food is left for it. If threatened, will roll into a ball to be protected by its spines. Hibernates. 16–26 cm.

Mole *Talpa europaea.* A very common animal, but absent from Ireland and the larger Scottish islands, except Mull and Skye. Spends all its life underground, so is rarely seen on the surface, but the molehills it throws up in the course of its burrowing are familiar, and often unwelcome, features of the countryside. Its diet consists largely of earthworms. 11–16 cm.

Pigmy Shrew *Sorex minutus.* The smallest native mammal in Britain and Ireland, widespread and frequent throughout both, including all the major islands except the Scillies, Shetlands, and the northernmost Orkneys. Only 5.7 cm long, the Pigmy Shrew can readily be differentiated from a young Common Shrew by its tail, which is always at least two-thirds as long as the head and body combined.

Common Shrew *Sorex araneus.* Generally commoner than the Pigmy Shrew, but absent from all Ireland, the Isle of Man, the Outer Hebrides and the northern isles. Frequents all kinds of rural habitats, such as grassland, scrub and moorland. Usually about 7.6 cm long, and its tail is only half as long as the head and body combined.

1

Water Shrew *Neomys fodiens* (1). Widespread but rather local in England, Wales and S Scotland, thinning out northwards, and absent from Ireland and most islands. Unlike other shrews it is semi-aquatic, and frequents the margins of streams and ponds, being very sensitive to pollution. Larger than other shrews, up to 10.2 cm long, the Water Shrew has a keel of hairs underneath its tail, which acts as a rudder when it swims. It may or may not have whitish underparts.

uni-coloured form of 1

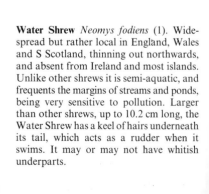

RODENTS Rodentia

Dormouse *Muscardinus avellanarius*. Confined to England and Wales, and now rather uncommon, except S of the Thames, it inhabits woods, especially where there is a Hazel understorey and some Honeysuckle. Dormice feed on hazelnuts and use Honeysuckle bark to build their nests, usually in a tree. They hibernate in a nest on the ground. Distinguishable from other rodents by their furry tails. 12.5–15 cm.

Fat Dormouse *Glis glis*. This Continental species, larger than the native Dormouse, was released at Tring, Hertfordshire, in 1902, and has established itself in a nearby area of the Chilterns. It was greatly relished by Roman epicures, and was fattened in captivity before being served with a sweet sauce of honey and poppy seeds. 25–32 cm.

Bank Vole *Clethrionomys glareolus*. A common small rodent of woodland, scrub and hedgerows, which occasionally climbs up twigs after wild fruits. It occurs throughout the British mainland, but is absent from most of the islands (except Anglesey and the Isle of Wight) and from all but the very SW of Ireland. It has more rufous fur and a longer tail than the Field Vole (below). 9–11 cm.

Water Vole *Arvicola terrestris*. This is the animal that makes a plop as you walk along a stream bank, and can then be seen swimming rapidly away. It occurs in and by fresh water throughout England, Wales and S Scotland, but in the Highlands is common only in the E, and is absent from Ireland. In N Scotland its fur may be black. 29–32 cm.

Field Vole or **Short-tailed Vole** *Microtus agrestis*. One of the 2 commonest small rodents of the countryside, found throughout the British mainland, but not in Ireland, the Isle of Man or the northern isles. It prefers more open country than the Bank Vole, and differs in its greyer-brown fur and very short tail. Vole populations are liable to increase rapidly and produce 'plagues'. 12–15 cm.

Orkney Vole *Microtus arvalis*. Replaces the Field Vole in both Orkney, where it is believed to have been introduced by the Norsemen 1,000 years ago, and in Guernsey in the Channel Isles. Slightly larger than the Field Vole, it has browner fur and a longer tail, but not as long as the Bank Vole's. 13–16 cm.

Harvest Mouse *Micromys minutus*. Half the 11.5 cm length of Britain's smallest rodent consists of its long, prehensile tail; it is also our most rufous mouse. Commonest S of the Trent, it is local and rare in Wales and the N and absent from Ireland. It makes breeding nests like a cricket ball, slung between 2 corn-, grass- or reed-stalks, and sleeping nests of dried grass on the ground.

Yellow-necked Mouse *Apodemus flavicollis*. This larger, more brightly coloured relative of the common Wood Mouse occurs in similar habitats in 2 main areas, SE England below a line from Suffolk to Dorset, and the Severn valley in the W. It has a yellowish spot on its chest. Like the Wood Mouse, it may enter houses in the winter. 18–20 cm.

Wood Mouse or **Long-tailed Field Mouse** *Apodemus sylvaticus*. The commonest mouse of the countryside, found in woods, scrub and hedgerows throughout Britain and Ireland, including all the larger islands. Its fur is browner than the Yellow-necked Mouse's and it also lacks its larger relative's yellow chest spot. 17–19 cm.

House Mouse *Mus musculus*. The most familiar British and Irish wild animal, entering almost every house at some time or other, but also frequent in the countryside, especially in and near corn ricks (a rapidly diminishing habitat). It is greyer-brown than the Wood Mouse, which may also enter houses, and has much less pale underparts. 16–18 cm.

Black Rat or **Ship Rat** *Rattus rattus*. Both British rats are aliens, the Black Rat having arrived first, in the Middle Ages, and been largely displaced by the Brown Rat (below), which came about 250 years ago. The Black Rat is now largely confined to seaports. It can be told from the Brown Rat by its slenderer appearance, sharper muzzle and longer tail; its fur may be either black or brown. 45–55 cm.

Brown Rat or **Common Rat** *Rattus norvegicus*. Much the commoner of our 2 rats, and found everywhere, in towns and villages and in the farming countryside; especially frequent near rubbish dumps. It is always brown and has a blunter muzzle and shorter, stouter tail than the Black Rat. It is one of the few really undesirable wild animals, because of the economic damage it does. 35–55 cm.

Red Squirrel *Sciurus vulgaris*. Squirrels are specialised tree-living rodents. The native Red Squirrel can be told by its rufous fur and characteristic ear-tufts. Once found in woodland all over Britain and Ireland, it is now very local in England and commonest in the Highlands. It can be a problem to the forester when it nips off the leading shoots of young trees. 28–40 cm.

Grey Squirrel *Sciurus carolinensis*. An invader from N America, let loose by 19th-century landowners, it is now much commoner than the Red Squirrel in England and Wales, but more local in Scotland and Ireland. It is always greyer than the Red Squirrel but can look quite rufous, and never has ear-tufts. It is an even more serious forestry pest, especially when it strips bark off trees. 45–50 cm.

Coypu *Myocastor coypus*. Britain's largest rodent is an alien from S America, imported to stock fur farms (it is called Nutria in the fur trade) and allowed to escape. It has established itself firmly in E Anglia, especially the Norfolk Broads, where it damages the river banks by burrowing and by overgrazing the marsh plants. At one time it had a wider distribution, but has been controlled. 95–100 cm.

RABBITS AND HARES   Lagomorpha
Rabbit *Oryctolagus cuniculus*. Still the most likely wild animal to be seen on a country walk anywhere in Britain and Ireland, having largely recovered from the great myxomatosis epidemic of the 1950s. Rabbits are not native, but were introduced in the Middle Ages, probably in the late 12th century, for their meat and fur, then a great luxury. 35–40 cm.

Brown Hare *Lepus capensis*. Widespread and frequent in farmland throughout the British mainland, but local and rare in the W Highlands; absent from Ireland. Substantially larger than a Rabbit, with warmer brown fur and longer ears. In early spring the males have a habit of chasing and sparring that has earned them the name of 'mad March hares'. 60–70 cm.

Blue Hare or **Mountain Hare** *Lepus timidus*. So called because of its blue–grey fur and mountain and moorland habitat. Also differs from the Brown Hare by its smaller size and shorter ears, and by turning white in winter. The only hare in Ireland, it is much the commoner of the 2 in the Scottish Highlands, and also occurs locally in S Scotland, the S Pennines and Snowdonia. 50–65 cm.

S   W

CARNIVORES Carnivora

Stoat *Mustela erminea*. Found throughout Great Britain, except for some of the Scottish islands, and Ireland, where it is confusingly called a Weasel. Substantially larger than the Weasel, from which it can always be told by the black tip to its tail, and in winter in the N by its white fur. White Stoats with their black-tipped tails provide the fur called ermine. 35–40 cm.

Weasel *Mustela nivalis*. Smaller than the Stoat, with no black tip to its tail, and never turning white in winter. Widespread and common on the mainland of Great Britain, but absent from Ireland and most of the islands, except Anglesey, Skye and the Isle of Wight. Preys on rats, mice, voles and young Rabbits, but nevertheless, like the Stoat, still persecuted by misguided gamekeepers. 20–25 cm.

Polecat *Mustela putorius*. Once widespread and common, but persecution by farmers and gamekeepers because it preys on their poultry and gamebirds has restricted it to Wales and some of the border counties. Elsewhere reports of this outsize dark brown Stoat are likely to be of escaped Polecatferrets, the dark form of the closely related domestic Ferret. 45–55 cm.

American Mink *Mustela vison*. A fur-farm escape, which has established itself since the 1950s on many rivers throughout Britain and Ireland, where it preys on fish and water-fowl. It is Polecat-sized, and all-dark with a white spot on its chin, and is not often found away from fresh water or marshes. Otters get blamed for many of its misdeeds. 42–65 cm.

Pine Marten *Martes martes*. More attractive than the somewhat smaller Polecat, with the whole throat and chest pale cream, instead of just the muzzle whitish. Once widespread throughout Britain, it is now only at all frequent in the NW Highlands, with scattered outposts in N England and N Wales. It occurs, however, throughout Ireland. Unpopular with gamekeepers. 65–75 cm.

Otter *Lutra lutra*. Now one of our most endangered mammals everywhere, except in W Scotland and Ireland. A fish-eater, it has been persecuted by fishermen and has suffered from polluted water and prey, but is now protected in England and Wales. Much larger than the Mink, which is now more likely to be seen in the S, it has a longer tail and a white throat. 95–130 cm.

Wild Cat *Felis sylvestris*. Britain's only wild member of the Cat Family bears a close resemblance to the domestic tabby, but is more striped, with a bushier, blunt-ended tail marked with black rings. Confined to the Scottish Highlands, it is gradually increasing its range as the pressure of gamekeeping relaxes. 'Wild cats' that are seen anywhere else are always feral domestic cats. 75–95 cm.

Fox *Vulpes vulpes*. Since the final disappearance of the Wolf some 200 years ago, the Fox has been Britain's only wild member of the Dog Family. Widespread and common in both Britain and Ireland, except for the Isle of Man and most of the larger Scottish islands. Increasingly present in towns and suburbs, where it is not persecuted. Fairly omnivorous. 100–110 cm.

Badger *Meles meles*. Commoner in wooded districts than is usually realised, it is present throughout the British and Irish mainlands, also on Anglesey and the Isle of Wight. Badger-watching is an increasingly popular pastime, but illegal Badger-digging still occurs in some districts. Badgers are more or less omnivorous and very rarely do any real harm to the farmer. 80–95 cm.

PINNIPEDIA   Seals
Common Seal *Phoca vitulina*. The smaller of Britain's 2 native seals, it is distinguished also by its shorter muzzle and more domed head. It occurs all round the coasts of Scotland and Ireland, and all down the E coast of England, breeding as near to London as sandbanks in the Wash (the largest British colony) and on the Essex coast. 140–180 cm.

Grey Seal *Halichoerus grypus*. The larger of the 2 British seals, and the more widespread, frequenting all coasts except those of SE England, but breeding mainly in the N and W, especially on the Farne Islands off Northumberland and some of the Scottish islands. Has a much longer muzzle and flatter head than the Common Seal. It is often seen as a curious head and neck watching humans on land from a safe distance out at sea. Grey Seals spend much of their life at sea and only really come to land in the autumn breeding season. 200–240 cm.

BATS Chiroptera

Lesser Horseshoe Bat *Rhinolophus hipposideros* (1). The 2 horseshoe bats are distinguished by their extraordinary horseshoe-shaped nose-leaves and the lack of a small, pointed inner ear or tragus. Both roost in caves. The Lesser is found in Wales and some western parts of England and Ireland. WS 22–25 cm.

Greater Horseshoe Bat *Rhinolophus ferrumequinum*, found only in SW England and S Wales, is the most endangered British bat. WS 34–39 cm.

Daubenton's Bat *Myotis daubentoni* (2). Sometimes known as the Water Bat because it feeds over fresh water, though other bats feed here too and Daubenton's Bat also feeds away from water. Found throughout Britain and Ireland, but scarce in SW Ireland and absent from the far N of Scotland. Intermediate in size between the large Noctule and the small Pipistrelle, (opposite), Daubenton's, the Whiskered (below) and Natterer's (below) Bats are hard to distinguish on the wing. WS 23–27 cm.

Whiskered Bat *Myotis mystacinus* (3). This bat occurs throughout Ireland, in England and Wales S of the Humber, and more locally in N England and S Scotland. It roosts in caves and buildings, and sometimes flies by day. WS 21–24 cm.

Natterer's Bat *Myotis nattereri* (4). Another medium-sized bat that is frequent throughout England, Wales and Ireland and occurs in Scotland as far N as the southern Highlands. Hawks for insects both over water and around trees; like all bats, it feeds almost entirely on flying insects, but also picks a few insects off the leaves of trees. WS 25–30 cm.

Long-eared Bat *Plecotus auritus* (1). One of the few bats that are comparatively easy to identify, if its long ears can be seen. The only other bats with long ears are both rare. The Long-eared Bat is common or frequent throughout Britain and Ireland, except for the far N of Scotland and the Scottish islands apart from Skye. Roosts in trees and buildings. WS 23–28 cm.

Noctule *Nyctalus noctula* (2). One of our 3 large bats, widespread and common in England and Wales, but more local in S Scotland, and absent from N Scotland, Ireland and all the major islands. It often emerges before sunset, when its golden brown fur can be seen in a good light. WS 32–39 cm. The dark brown **Serotine** *Eptesicus serotinus* is equally large, but confined to SE England.

Pipistrelle *Pipistrellus pipistrellus* (3). This tiny species is our smallest bat, and probably our commonest, occurring throughout Britain and Ireland and on all the major offshore islands except Shetland. It may be seen flying in many kinds of habitat, often over water; usually emerges about 20 minutes after sunset, but occasionally appears in the daytime. WS 19–25 cm.

EVEN-TOED UNGULATES Artiodactyla
Feral Goat *Capra hircus*. Goats which have escaped from domestication from the Middle Ages onwards have established themselves as feral breeding populations in many parts of Highland Britain and Ireland, including Snowdonia and the Cheviots. They may be almost any colour or combination of colours: black, white, grey, dark brown, light brown, piebald or skewbald.

Muntjac or **Barking Deer** *Muntiacus reevesi*. A small deer, the size of a large dog, which originally escaped from Woburn Park in Bedfordshire, and has now spread over the woodlands of a large area of S and Midland England, being especially common in the Chilterns. Males have small antlers and distinct tusks; females' tusks are shorter. SH 45–48 cm.

Fallow Deer *Dama dama*. Generally the commonest deer in the woodlands of England, Wales and Ireland; rather uncommon in Scotland. Males (bucks) are distinguished by their flattened antlers, and both sexes have white spots and look more rufous in summer. Fallow Deer are not native, and are believed to have been introduced by the Normans. SH 85–95 cm.

Red Deer *Cervus elaphus*. Our largest native deer and largest land mammal, widespread and locally common in the Scottish Highlands; elsewhere most frequent in SW Scotland, the Lake District, SW England and the Breckland of E Anglia; very local in Ireland. Males (stags) grow a fine pair of antlers between April and July, shedding them again in February. SH 115–120 cm.

Sika Deer *Cervus nippon*. Another introduced deer, present in widely scattered places in England, Scotland and Ireland, especially the New Forest in Hampshire and Bowland Forest on the Yorkshire/Lancashire border. It has slenderer antlers than both Red and Fallow Deer, but is spotted in summer like the Fallow. Antlers are not cast until April. SH 75–80 cm.

Roe Deer *Capreolus capreolus*. The smaller of our 2 native deer is widespread and common in Scotland and N England, but elsewhere only at all common in Hampshire and adjoining counties and in parts of E Anglia. Intermediate between Fallow and Muntjac in size, it is never spotted, and has much simpler antlers than either Red or Fallow, shedding them in November. SH *c*. 65 cm.

144

Mammal Tracks

One of the best ways of discovering what mammals, and for that matter birds and amphibians too, are about is to look out for their footprints and learn to identify them. After a snowfall or during very muddy conditions are the best times to look.

A good start is to count the number of toes, as indicated below:

1 toe (hoof): domestic horse or donkey.

2 toes (cloven hoof): deer, cattle, sheep, goats, pigs.

4 toes, with claws: Fox, Wild Cat, domestic dog and cat.

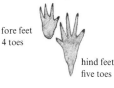

fore feet 4 toes

hind feet five toes

Fore feet 4-toed, hind feet 5-toed, with claws: rodents (rats, mice, voles, squirrels).

5 toes, with claws: Hedgehog, Mole, shrews, Badger, Otter, Stoat, Weasel, Pine Marten, Polecat.

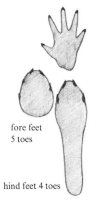

fore feet 5 toes

hind feet 4 toes

Fore feet 5-toed, hind feet 4-toed, with claws: Rabbit, hares.

The gait of a mammal, whether it is walking, trotting, galloping, hopping or jumping, can also often be deduced from a study of tracks. In a walking gait the animal goes right foreleg, left hindleg, left foreleg, right hindleg. As it accelerates to a trot, the 2 pairs right foreleg/left hindleg and left foreleg/right hindleg tend to leave the ground together. By the time a gallop is reached, much of the time is spent with all 4 legs in the air. In hopping and jumping, the limbs are moved in a different pairing, right fore/left fore, and right hind/left hind. All these can be observed from a clear set of footprints, though there are complications, as for instance when deer place their hind feet in the tracks made by the fore feet, and Foxes and cats arrange their steps in such a way that they make what is virtually a straight line.

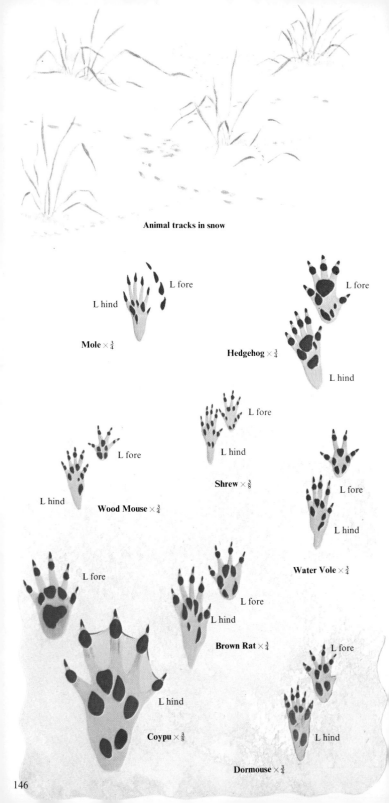

Animal tracks in snow

Mole $\times \frac{3}{4}$

L hind
L fore

Hedgehog $\times \frac{3}{4}$

L fore
L hind

Wood Mouse $\times \frac{3}{4}$

L fore
L hind

Shrew $\times \frac{3}{8}$

L fore
L hind

Water Vole $\times \frac{3}{4}$

L fore
L hind

Brown Rat $\times \frac{3}{4}$

L fore
L hind

Coypu $\times \frac{3}{8}$

L fore
L hind

Dormouse $\times \frac{3}{4}$

L fore
L hind

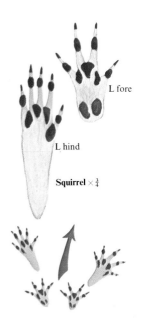

Squirrel × ¾

L fore

L hind

L fore

Rabbit × ¾

Squirrel's jumping track

Rabbit's track

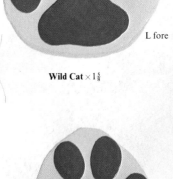

L fore

Wild Cat × 1⅝

L fore

Hare × ¾

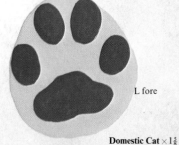

L fore

Domestic Cat × 1⅝

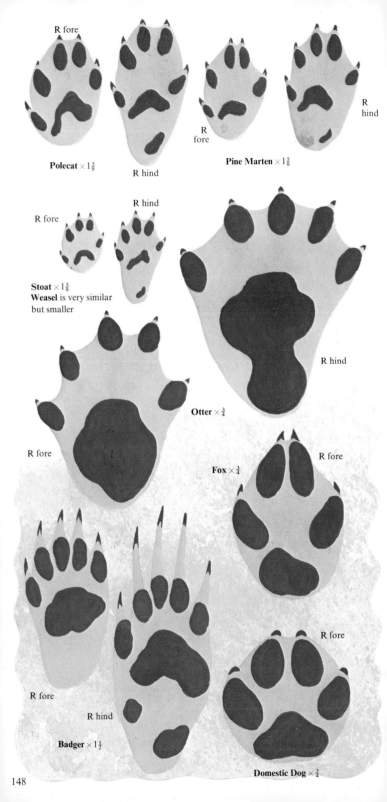

R fore

Polecat × 1⅝

R hind

R fore

Pine Marten × 1⅝

R hind

R fore

R hind

Stoat × 1⅝
Weasel is very similar
but smaller

R hind

Otter × ¾

R fore

Fox × ¾

R fore

R fore

R fore

R hind

Badger × 1½

R fore

Domestic Dog × ¾

148

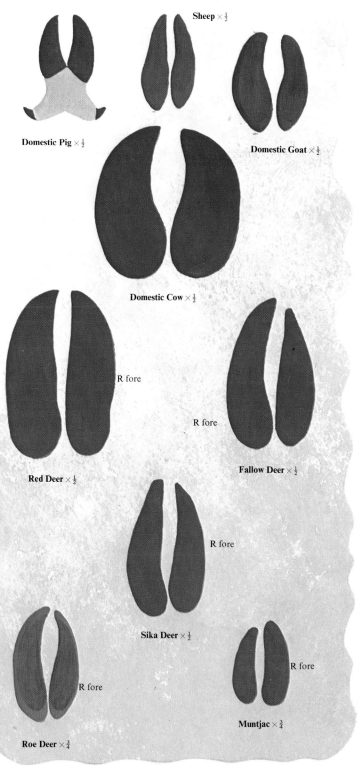

Sheep $\times \frac{1}{2}$

Domestic Pig $\times \frac{1}{2}$

Domestic Goat $\times \frac{1}{2}$

Domestic Cow $\times \frac{1}{2}$

Red Deer $\times \frac{1}{2}$

R fore

Fallow Deer $\times \frac{1}{2}$

R fore

R fore

Sika Deer $\times \frac{1}{2}$

R fore

Roe Deer $\times \frac{3}{4}$

R fore

Muntjac $\times \frac{3}{4}$

R fore

Birds

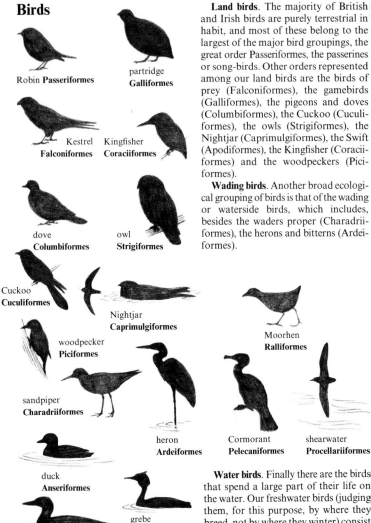

Robin **Passeriformes**

partridge **Galliformes**

Kestrel **Falconiformes**

Kingfisher **Coraciiformes**

dove **Columbiformes**

owl **Strigiformes**

Cuckoo **Cuculiformes**

Nightjar **Caprimulgiformes**

Moorhen **Ralliformes**

woodpecker **Piciformes**

sandpiper **Charadriiformes**

heron **Ardeiformes**

Cormorant **Pelecaniformes**

shearwater **Procellariiformes**

duck **Anseriformes**

grebe **Podicipitiformes**

diver **Gaviiformes**

Land birds. The majority of British and Irish birds are purely terrestrial in habit, and most of these belong to the largest of the major bird groupings, the great order Passeriformes, the passerines or song-birds. Other orders represented among our land birds are the birds of prey (Falconiformes), the gamebirds (Galliformes), the pigeons and doves (Columbiformes), the Cuckoo (Cuculiformes), the owls (Strigiformes), the Nightjar (Caprimulgiformes), the Swift (Apodiformes), the Kingfisher (Coraciiformes) and the woodpeckers (Piciformes).

Wading birds. Another broad ecological grouping of birds is that of the wading or waterside birds, which includes, besides the waders proper (Charadriiformes), the herons and bitterns (Ardeiformes).

Birds are a highly specialised group of warm-blooded vertebrates, which maintain a constant temperature and whose forelimbs have developed into wings, which together with their other highly distinctive character, feathers, enable them to fly. Some 200 species breed in Britain and Ireland, of which about 50 are summer visitors only. Approximately 100 more species are regularly winter visitors, or passage migrants *en route* to or from their more northern breeding areas and their more southern winter quarters. There are also some 200 stragglers of varying degrees of rarity.

Water birds. Finally there are the birds that spend a large part of their life on the water. Our freshwater birds (judging them, for this purpose, by where they breed, not by where they winter) consist of the ducks, geese and swans (Anseriformes), the divers (Gaviiformes), the grebes (Podicipitiformes) and the Moorhen and Coot (Ralliformes). Our seabirds proper, i.e. those that breed by the sea and spend most of their lives on it, include the petrels and shearwaters (Procellariiformes), the Gannet, Cormorant and Shag (Pelecaniformes), and the gulls (Charadriiformes).

Of course there are some birds that cross even these very broad groupings. The wagtails, for instance, though song-birds, are partially waterside birds; the Corncrake, though a rail, is a land bird, and some plovers, other waders and gulls breed on quite dry land, often far from any substantial area of water.

DIVER FAMILY Gaviidae

Great Northern Diver *Gavia immer*. Like other divers is well adapted to aquatic life and rarely comes to land; has a loud, wailing cry. Its stouter bill and more peaked forehead distinguish it from the smaller Black-throated at all times; its throat pattern is distinctive in spring and summer. A winter visitor, most frequent in coastal waters of N and E; has bred. 70–90 cm.

Black-throated Diver *Gavia arctica*. Breeds on larger lochs in the W half of Scotland; more widespread in coastal waters in winter. Like other divers', its nest is located very close to loch bank. In breeding plumage it differs from the larger Great Northern in its more conspicuously black throat; at other times has a less stout bill and less peaked forehead. 60–70 cm.

Red-throated Diver *Gavia stellata*. Our commonest and smallest diver, breeding on smaller lochs and tarns in W Scotland, Orkney, Shetland and NW Ireland; more widespread round coast in winter. Easily told by its red throat in spring and summer, and in winter by its smaller size and uptilted bill. Has a rather quacking flight note and a wail likened to a cat miaowing. 55–65 cm.

GREBE FAMILY Podicipedidae

Great Crested Grebe *Podiceps cristatus* (1). Our largest grebe, very distinctive in breeding plumage with its crest and tippets, used in its remarkable head-wagging courtship display. Shows a whitish wing-bar in flight. Breeds on ponds and lakes, except in NW Scotland; also frequents coastal waters in winter. Nest always over water, often anchored to aquatic vegetation. 46–51 cm.
Red-necked Grebe *P. grisegena* (2) differs from Great Crested in yellow and black beak and no white eye-stripe. 42 cm.

Slavonian Grebe *Podiceps auritus* (3). A scarce breeder in the Scottish Highlands, more widespread in coastal waters in winter, when it is best told from the Black-necked by its straight bill and distinctive face pattern. 31–38 cm. **Black-necked Grebe** *P. nigricollis* (4) is rarer but more widespread as a breeder; commoner in coastal waters in winter than Slavonian, from which it differs in its uptilted bill and less white on face. 28–34 cm.

Little Grebe or **Dabchick** *Tachybaptus ruficollis*. Our smallest and commonest grebe, easily identified in breeding plumage; in winter best told by lack of white on its face. Breeds on lakes, ponds and slow-flowing streams; also in coastal waters in winter. Like all grebes it is an accomplished diver. It has a whinnying cry. The nest is a mass of floating vegetation. 25–29 cm.

SHEARWATER FAMILY Procellariidae
Fulmar *Fulmarus glacialis*. Gull-like in plumage, it is readily distinguished by its stiff-winged, shearwater-like flight, and at close quarters also by its tubular nostrils, characteristic of all shearwaters and petrels. Now breeds on virtually all cliffed coasts of British Isles, having spread from St Kilda alone in the past 100 years. In defence of nest will spit foul-smelling oil at intruders. 45–50 cm.

Great Shearwater *Puffinus gravis* (1) and **Sooty Shearwater** *P. griseus* (2) are uncommon visitors, usually in autumn, to the seas to the N and W of both Scotland and Ireland, occasionally seen a mile or two offshore. 46 cm and 41 cm respectively.

Manx Shearwater *Puffinus puffinus*. Much the commonest shearwater of British seas, easily told by its characteristic flight, skimming stiff-winged over the waves, alternately showing its dark upperparts and pale underparts. Breeds in burrows on marine islands off the W coast from Scillies to Shetland, and round the Irish coast. The browner Mediterranean race *balearicus* is an uncommon visitor to the English Channel. 30–38 cm.

Storm Petrel *Hydrobates pelagicus*. Britain's smallest seabird, the original Mother Carey's chicken, and well known for following ships with its direct but rather weak and fluttering flight. White rump as conspicuous as a House Martin's. Breeds on offshore islands on W coasts of both Britain and Ireland. Can be detected in nesting burrows by both churring call and pungent smell. 14–18 cm.

Leach's Petrel *Oceanodroma leucorhoa*. One of the rarest British breeding birds, confined to remote marine islands from St Kilda to Foula, and formerly off W Ireland. It breeds in burrows or among boulders. The forked tail is hard to see in flight, which, however, is characteristically buoyant and darting to and fro. Unlike the Storm Petrel (above) it does not follow ships. 19–22 cm.

BOOBY FAMILY Sulidae
Gannet *Sula bassana*. Britain's largest seabird; adults are easily told by contrast between white body and black wing-tips and habit of diving from a height. Dark brown immatures, speckled whiter with each moult, are more confusing. Breeds at a dozen colonies round the N and W coasts of Britain, from Pembrokeshire to Flamborough Head, and 3 colonies off S Ireland. The most famous gannetries are at Bass Rock in the Firth of Forth and Ailsa Craig in the Firth of Clyde. 87–100 cm.

CORMORANT FAMILY
Phalacrocoracidae

Cormorant *Phalacrocorax carbo*. The largest all-dark British seabird, with a white face-patch and in breeding season also white thigh-patches. Young birds are whitish beneath. On water holds bill at a pronounced upward angle, unlike divers and grebes. Breeds on cliffs and rocky islets on all coasts except between Yorkshire and Isle of Wight. In winter also on low-lying eastern coasts and at fresh water inland. 80–100 cm.

Shag *Phalacrocorax aristotelis*. Smaller than Cormorant and with greenish instead of bluish sheen; also lacks any white patches and has recurved crest in breeding season. Young birds are whitish mainly on throat. Same breeding range as cormorant, but commoner in N and W. Uncommon off SE coasts in winter, and rare inland. Flies with faster wing-beats than Cormorant. 65–80 cm.

HERON FAMILY   Ardeidae

Grey Heron *Ardea cinerea* (1). The largest long-legged British bird that wades in water, often seen also as a huge, broad-winged bird flapping slowly overhead. Breeds throughout British Isles, except Shetland, usually in trees, but occasionally in reed-beds or on cliffs. 90–98 cm. **Purple Heron** *A. purpurea* (2) is a rare visitor from the Continent, told by its striped rufous neck and chestnut breast. 78–90 cm.

Bittern *Botaurus stellaris*. Much more often heard, with its low booming call, like a distant lowing cow, than seen, but occasionally flies low over the extensive reed-beds which are its main habitat. An uncommon breeder, almost confined to E England from the Humber to Kent, and mainly in E Anglia. In winter, especially in hard weather, it is a little more widespread. 70–80 cm.

IBIS FAMILY   Threskiornithidae

Spoonbill *Platalea leucorodia*. A large, white, long-legged bird, unmistakable when its large, dark, spoon-shaped bill can be seen. Adult has buff patch on breast; young birds have pinkish bills and black wing-tips and lack the adult's crest. An uncommon visitor, chiefly to E coast estuaries in late summer and autumn, probably stragglers from Dutch breeding colonies; also in winter in SW England and S Ireland. 80–90 cm.

153

DUCK FAMILY Anatidae

Mute Swan *Cygnus olor*. Much the commonest swan, and the only one likely to be seen in spring and summer. The orange bill and black knob, smaller in the female, are distinctive. Neck usually held in a graceful curve. Normally silent, but will hiss or snort when annoyed, and its wings make a fine, throbbing *hompa, hompa* sound in flight. Lakes, ponds, rivers, estuaries. Nest a huge pile of vegetation. 152 cm.

Whooper Swan *Cygnus cygnus*. The larger of the 2 yellow-billed winter visitor swans, with the yellow extending more than halfway to the tip of the knobless bill. Neck usually held upright. Has a loud, goose-like clanging call-note, and its wings whistle rather than throb in flight. Mainly in the N, and has bred in Scotland. 152 cm.

Bewick's Swan *Cygnus columbianus bewickii*. Now considered to be a race of the N American Whistling Swan. Smaller than Whooper Swan, it has shorter neck and much less yellow on its bill. Voice is more musical. Mainly in the S, notably at Wildfowl Trust at Slimbridge, Glos. As in all swans, the immature in greyish. 122 cm.

Brent Goose *Branta bernicla*. The smallest of our 3 black-necked geese, and the only one with an all-black head. Highly gregarious winter visitor, usually on estuarine and other mud-flats, especially where its favourite Eel-grass *Zostera* grows. Mainly in the E, migrating from Arctic Europe; also in Ireland, coming from Greenland and Spitzbergen. 56–61 cm.

Barnacle Goose *Branta leucopsis*. The smaller of our 2 black-necked geese, it has white on its face. Highly gregarious – the flocks' collective cries sound like yapping dogs. A winter visitor to grassland and marshes by the sea in Scotland and Ireland. It not infrequently escapes from collections of ornamental waterfowl inland. 58–69 cm.

Canada Goose *Branta canadensis*. The largest of our 3 black-necked geese, and much the most likely to be seen inland. A long-established introduction from N America, it now breeds widely by fresh water almost throughout Great Britain. In winter it may resort to coastal waters, especially Beauly Firth, N Scotland. The call is a loud double trumpeting, *ker konk*. 92–102 cm.

Greylag Goose *Anser anser*. The largest and heaviest of British grey (actually grey–brown) geese, best distinguished by its pale grey forewing in flight and its combination of orange bill and pink legs. A few native birds still breed in N Scotland and there are some introduced flocks elsewhere. As a winter visitor it frequents northern farmland and coastal marshes. Voice not unlike that of the farmyard goose, whose ancestor it is. 75–90 cm.

Pink-footed Goose *Anser brachyrhynchus*. The smallest of the common grey geese, best told by its dark head and neck, comparatively buoyant flight and combination of pink bill and pink legs. Winter visitor only, from Iceland and Spitzbergen to farmland and coastal marshes in S Scotland and N and E England. Its calls are more high-pitched than those of other grey geese. 60–75 cm.

Bean Goose *Anser fabalis*. Much the most local of Britain's 4 regularly wintering grey geese, best distinguished by its heavy, Greylag-like build, dark head and neck, and combination of orange bill and orange legs. Lacks the Greylag's pale forewing. Winter visitor to coastal marshes, mainly in SW and central Scotland. 66–84 cm.

White-fronted Goose *Anser albifrons*. The white on the face and heavily barred underparts make this the easiest of the grey geese to identify at close range, but immatures need to be distinguished from Pink-footed by their orange legs. The bill is pink in the European race (1) wintering in England, but orange–yellow in the Greenland race (2) in Scotland and Ireland. This is the wild goose on the marshes at the Wildfowl Trust, Slimbridge, Glos. 65–78 cm.

Shelduck *Tadorna tadorna*. A large black and white duck, resembling a small goose, common all round the coast on sandy and muddy seashores, nesting in burrows, sometimes even on heaths a few miles inland. Drake distinguished by red knob on bill; immatures rather confusingly different from adults, with no chestnut in plumage and some white on face. 58–67 cm.

Mandarin Duck *Aix galericulata*. An introduced species, native to China and Japan, which has established itself in parts of S England where there are ponds or lakes in woodland, especially in and around Windsor Great Park, Berkshire. It nests in holes in trees. Drake strikingly different from duck. 41–49 cm.

155

Mallard *Anas platyrhynchos*. Much the largest and commonest British wild duck, both inland and on the coast, often nesting by quite small ponds on moorland and in farmland. Drake unmistakable with green head, white neck-ring and purple–brown breast. Females can always be told from other brown ducks by blue and white speculum (bar on wing). Voice a quack like its descendant, the farmyard duck. 50–65 cm.

Gadwall *Anas strepera*. One of the rarer British dabbling ducks (ducks which normally do not dive), mainly in Scotland and E Anglia. Drake has distinctive grey plumage, but ducks needs to be distinguished from Mallard duck by black and white speculum. Both have more peaked head and more delicate bill than Mallard. Duck's quack is softer. 46–56 cm.

Wigeon *Anas penelope*. Mainly a winter visitor, whose presence can often be detected on a fog-bound reservoir or estuary by its characteristic *whee-oo* call. Breeds most commonly in Scotland. Drake's pale forehead differentiates it from drake Pochard (a diving duck), and duck's peaked forehead and green speculum from female Mallard. Both can readily be picked out in flight by their conspicuous white wing-bar. 45–57 cm.

Shoveler *Anas clypeata*. A duck swimming along with its beak in the water is likely to be a Shoveler, whose spoon-shaped bill is specially designed for trawling the surface. Bill and blue forewing both separate Shovelers from other ducks; drake's chestnut flanks are also distinctive. Much commoner in autumn and winter than in breeding season, when seen mainly in the E. 44–52 cm.

Teal *Anas crecca*. Britain's smallest dabbling duck, often distinguishable at a distance by its agile flight as it springs from the water. Drake's head pattern is very distinctive, and under-tail pattern often enables species to be picked out in poor light. Duck has green speculum. Commoner in autumn and winter than in breeding season, when it is most frequent in hill districts. 34–38 cm.

Garganey *Anas querquedula*. The only summer visitor and much the least common British dabbling duck, being largely confined to SE England and E Anglia. Drake's white head-streak is distinctive, and duck can be told from smaller duck Teal by blue fore-wing. Drake's spring call has been likened to a single match rattling in a matchbox. 37–41 cm.

Pintail *Anas acuta*. Drake is in many ways the most elegant and handsome British duck, and quite unmistakable with its long tail. Female can be confused with duck Wigeon, but has a more pointed tail and a bronzy speculum. Fairly common in winter inland and on estuaries, but breeds only in a few widely scattered localities, from Orkney to Kent, by fresh and brackish water. 51–66 cm.

Tufted Duck *Aythya fuligula*. The commonest diving duck on southern park lakes and reservoirs, breeding by fresh water almost throughout the British Isles. Drake's head-tuft and dark back distinguish it from large drake Scaup; white at base of some ducks' bills is never as conspicuous as duck Scaup's. 40–47 cm.

Scaup *Aythya marila*. Slightly larger than Tufted Duck, from which the drake's grey back and lack of head-tuft and duck's large white face-patch both distinguish it. A winter visitor, mainly to N and E coastal waters, it has only rarely nested, in the Scottish Highlands. The Firth of Forth is a favoured locality. 42–51 cm.

Pochard *Aythya ferina*. Nowadays usually pronounced with a long 'o'. The other common diving duck of the S, it breeds less frequently and not so much in the N and W. Drake lacks the pale forehead of the drake Wigeon, which does not normally dive, and neither sex has the Wigeon's white forewing. 42–49 cm.

Eider *Somateria mollissima*. Drake's pattern of white above and black below is unmistakable, but immature and moulting birds can be very confusing. Sloping bill outline and barred breast distinguish female from duck Mallard. A resident duck of coastal waters, breeding widely in Scotland and S to Northumberland and Lancashire in England. Non-breeding birds penetrate further S still. 50–70 cm.

Common Scoter *Melanitta nigra* (1). Drake is the only British all-black duck, but other ducks can look black in poor light; the female's pale face is distinctive. At sea on most coasts in winter, breeding very locally in fresh water in Scotland and Ireland. 44–54 cm. **Velvet Scoter** *M. fusca* (2), with distinctive wing-bars, is a less common winter visitor, mainly in the E. 51–58 cm.

Goldeneye *Bucephala clangula*. A diving duck recognisable at a distance by its peaked head, and the drake at closer range by its white face-spot. A not uncommon winter visitor to both fresh and salt water, but breeds only very locally at one or two places in the Scottish Highlands. The Goldeneye's wings make a loud singing sound in flight. 42–50 cm.

Long-tailed Duck *Clangula hyemalis*. A winter visitor to coastal waters that has several times been suspected of breeding, and has probably done so in the northern isles. A diving duck, whose male and female and summer and winter plumages are both very different. Its very short bill is as good a field character as the drake's long tail. 40–47 cm.

Goosander *Mergus merganser*. A fresh-water diving duck that is a winter visitor to the S, but breeds in Scotland and N England and is spreading S to Wales. Drake's bulbous head and white breast and duck's short crest distinguish it from smaller Red-breasted Merganser, which shares its narrow 'saw' bill, serrated to enable it to catch fish. 58–66 cm.

Red-breasted Merganser *Mergus serrator*. Both sexes have distinctive tufted crests, and drake's dark breast also distinguishes it from the larger Goosander. Ducks and immatures of both species are popularly known as 'redheads'. Predominantly coastal in winter, the Red-breasted Merganser breeds widely in Scotland, Ireland, N England and Wales, mainly in the W, both inland and by the sea. 52–58 cm.

Smew *Mergus albellus*. The smallest of the 3 British sawbills is an increasingly scarce winter visitor, still as likely to be seen on reservoirs in the London area as anywhere. The striking drake, known to the old wildfowlers as the 'white nun', can look almost completely white at rest, but less so in flight. Dives more frequently than other sawbills. 38–44 cm.

Ruddy Duck *Oxyura jamaicensis*. This species is a N American diving duck that has recently established itself in Britain, thanks to escapes from Wildfowl Trust and other collections; it breeds mainly in the W from Somerset to Cheshire. The drake is unlike any British water-fowl, and the dumpy appearance of the pale-faced duck is also distinctive. 35–43 cm.

158

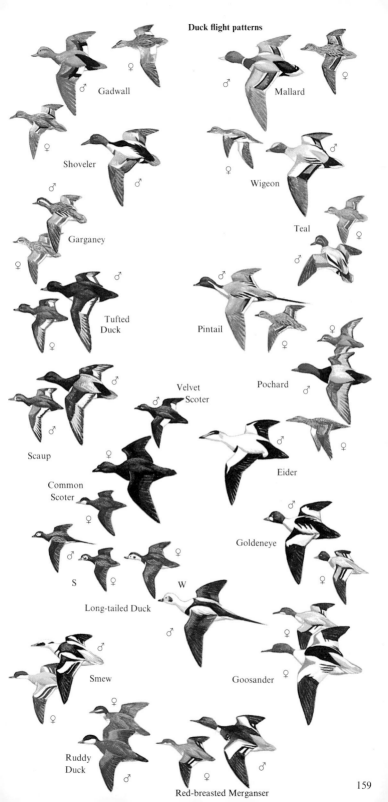

Duck flight patterns

Gadwall ♂ ♀

Mallard ♂ ♀

Shoveler ♀ ♂

Wigeon ♀ ♂

Garganey ♂ ♀

Teal ♂ ♀

Tufted Duck ♂ ♀

Pintail ♂ ♀

Scaup ♂ ♀

Pochard ♀ ♂ ♀

Velvet Scoter ♂ ♀

Common Scoter ♂ ♀

Eider ♂

Goldeneye ♂ ♀

Long-tailed Duck S ♂ ♀ W ♂ ♀

Smew ♂ ♀

Goosander ♀ ♀

Ruddy Duck ♂

Red-breasted Merganser ♀ ♂

159

OSPREY FAMILY Pandionidae

Osprey *Pandion haliaetus.* A lost Scottish breeding bird that has made a spectacular comeback in the past 20 years, with 20 or more pairs breeding in various parts of the Highlands, most notably at Loch Garten, Inverness-shire, and Loch of the Lowes, Perthshire. Pale head, white underparts and habit of plunging into water from a height are distinctive features. 55–58 cm.

HAWK FAMILY Accipitridae

Red Kite *Milvus milvus.* An almost lost British breeding bird that has also made a good comeback in its mid Welsh breeding haunts where there are now some 20–30 pairs. Forked tail, pale wing-patches, and at close range rufous plumage are points of distinction from the much commoner Buzzard. Frequents hill districts. 60–66 cm.

Sparrowhawk *Accipiter nisus* (1). Also returning since it has been protected from game preservers, and now quite widespread and not uncommon in wooded districts. Combination of long tail, blunt wings and 'flap and glide' flight distinguish it from the commoner Kestrel. 28–38 cm. The rare **Goshawk** *A. gentilis* (2) is an outsize Sparrowhawk. 48–62 cm.

Buzzard *Buteo buteo.* The largest commoner British bird of prey, now widespread in western hill districts, nesting in trees and on cliff ledges. Soars overhead, often half a dozen or more at a time, with broad wings like a giant moth, uttering a loud, gull-like *pee-oo* cry. Tail uniformly barred, with a dark bar at tip. 51–57 cm.

Honey Buzzard *Pernis apivorus.* One of the rarest British birds of prey, a summer visitor to a handful of southern and Midland woodlands. Distinguished from the Buzzard by a longer tail, with 2 dark bars at base beneath, and broader bar at tip. Has a well-known habit of raiding bees' and wasps' nests to feed on grubs. 52–60 cm.

Golden Eagle *Aquila chrysaetos.* Britain's largest and most majestic bird of prey, still quite frequent in the Scottish Highlands, much less so in SW Scotland, with one pair in the English Lake District. Immatures have whitish patches at base of tail and on each wing. The Golden Eagle's splayed-out primaries when soaring are a good field mark. 75–88 cm.

Marsh Harrier *Circus aeruginosus*. One of Britain's rarest breeding birds of prey, virtually confined to the marshlands of E Anglia, especially the Suffolk coast. Males are strikingly grey and brown when they fly, and females differ from our other harriers by their pale heads. Flight heavier and more Buzzard-like than other harriers'. Summer visitor only. 48–56 cm.

Hen Harrier *Circus cyaneus* (1). Now the commonest British harrier, breeding widely on moorlands in the N and W. Male, a striking blue–grey bird, lacks the black wing-bars of the now very rare and more southern **Montagu's Harrier** *C. pygargus* (2). Females of the 2 species are hard to separate, but Hen Harrier's white rump is more prominent. Immature Montagu's differs in its deep buff underparts. Hen 44–52 cm; Montagu's 43–47 cm.

FALCON FAMILY Falconidae

Peregrine *Falco peregrinus*. Has made a striking recovery from the days when it was being steadily exterminated by the pesticide residues in its prey, and is now restored to many parts of the N and W. A large bird with scimitar-shaped wings and a short tail, it swoops on its prey with a stupendous power dive. Female is appreciably larger than male. Nests on cliff ledges. 36–48 cm.

Hobby *Falco subbuteo*. A miniature Peregrine, with the same anchor-like outline, resembling a giant Swift, it is increasingly frequent in many wooded parts of the S. Often seen chasing Swallows and martins, its favourite prey. Dark moustachial streaks help to distinguish it from other small falcons. Nests in old crows' nests in trees. 30–36 cm.

Merlin *Falco columbarius*. Britain's smallest falcon, the male not much bigger than a Blackbird, is also beginning to recover from a pesticide-related population decrease. Breeds widely on the ground in northern moorlands, and frequents southern and eastern estuaries and marshes in winter. Lacks the Hobby's chestnut thighs and the Kestrel's rufous back. 25–30 cm.

Kestrel *Falco tinnunculus*. Generally the commonest British diurnal bird of prey, seen right in the middle of London and other cities. Best known for its habit of hovering in search of prey, usually small rodents. Rakish outline of long wings and tail is distinctive. Breeds throughout British Isles, except Shetland, in old crows' nests or on cliff ledges. 32–35 cm.

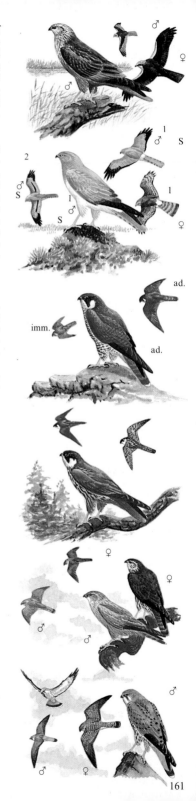

161

GROUSE FAMILY Tetraonidae

Red Grouse *Lagopus lagopus*. One of the staple gamebirds of British sportsmen, best known for its habit of flying up when disturbed on heather moorland, its sole habitat, with a loud cry of *gobak gobak* or *kok-kok-kok*. Confined to the N and W of the British Isles. The red wattle over the eye is visible only at close range. 37–42 cm.

Ptarmigan *Lagopus mutus*. The grouse of the high mountain-tops in the Scottish Highlands. In winter all-white, except for black tail, as camouflage on its snow-covered habitat, but in summer largely grey-brown above, turning greyer in autumn. Rarely found below 700 m. Main call a hoarse croak, *uk uk*. 34–36 cm.

Black Grouse *Tetrao tetrix*. Male (Blackcock) unmistakable with lyre-shaped tail, much smaller than Capercaillie. Female (Greyhen) and immatures differ from same-sized Red Grouse by narrow wing-bar and forked tail. A bird of the zone where woodland and moorland or heathland meet, well known for remarkable communal displays or leks in spring. Only in the N and W of Great Britain; not in Ireland. 40–55 cm.

Capercaille *Tetrao turogallus*. One of the largest British land birds, the size of an eagle. Male unmistakable; female and immatures differ from much smaller Greyhen by chestnut patch on breast, rounded tail and no wing-bar. Became extinct in Britain 200 years ago, but reintroduced in 1830s and now widespread and not uncommon in pinewoods in the Scottish Highlands. 60–87 cm.

PHEASANT FAMILY Phasianidae

Pheasant *Phasianus colchicus*. A well-known gamebird, not native but introduced by the Normans 900 years ago and now widespread and common, except in parts of NW Scotland. Cocks unmistakable, but may or may not have white neck-ring; hens distinguished from Partridges by size and long tail. Juveniles more confusable with Partridges but usually go with adults. 53–89 cm.

Quail *Coturnix coturnix*. Britain's smallest gamebird and the only summer visitor among them, most frequent in the S, and there mainly on chalk. Much more often heard, with characteristic *wet-mi-lips* or *quic-ic-ic* call, than seen. When flushed could be mistaken for juvenile Partridge which, however, usually goes with adult. 16–18 cm.

162

Red-legged Partridge *Alectoris rufa.* Introduced in the 18th century, it is now widespread and frequent on farmland, especially arable, in SE England. Larger than Partridge, and easily told by its barred flanks, black and white face pattern, and red legs. Call a loud *chuka chuka* or *chik chik chika*. 32–34 cm.

Partridge *Perdix perdix.* The most popular English gamebird, widespread and frequent, but thinning out northwards and westwards. Differs from Red-legged Partridge by its chestnut-coloured face-patch and horseshoe-shaped patch on breast. Call is a grating *keev* or *heev-it*. Mainly on farmland, but also heathland, shingle and sand-dunes. 29–31 cm.

RAIL FAMILY   Rallidae
Water Rail *Rallus aquaticus* (1). A shy, skulking bird, easily told by its long red bill and barred flanks when seen for more than a glimpse. Widespread but local in marshes and other wetlands with dense vegetation; commoner in winter and most often seen when forced into the open by hard weather. Voice very varied, clucking, grunting and squealing. 23–28 cm. The **Spotted Crake** *Porzana porzana* (2) is a rare visitor, even more rarely seen, as it is so skulking. 23 cm.

Corncrake *Crex crex.* Another extremely shy and skulking bird, much more often heard, uttering its harsh, grating *crex crex* call, than seen. When flushed, for instance by haymakers, it flies with dangling legs and shows the chestnut in its wing. Once widespread in farmland, the Corncrake has been decreasing for many years and is now only at all frequent in Ireland and the Hebrides. 27–30 cm.

Moorhen *Gallinula chloropus.* One of Britain's commonest and most widespread waterfowl, rare only in NW Scotland. Readily told by its red bill and forehead, white streak on flanks and white under tail coverts, which it frequently flicks. Juvenile lacks red on head and has pale throat; chick is all-dark. Often in town parks, and may feed on grassland away from water. Perches in trees and bushes. 32–35 cm.

Coot *Fulica atra.* Differs from Moorhen by its white bill and forehead and in flight by its white wing-bar; juvenile has more white on breast as well as throat, and chick has rufous head. Almost as widespread as Moorhen, but not on quite such small ponds. Aggressive chases are characteristic; it dives readily. 36–38 cm.

W   S

southern form

northern form

S

W

S

W

S

W

S

W

OYSTERCATCHER FAMILY
Haematopodidae

Oystercatcher *Haematopus ostralegus*. A very conspicuous, large black and white shore bird, with red bill and pink legs, that draws attention to itself with loud *kleep* cries as it flies offshore, and also by its 'piping parties' of several birds in mutual display. Breeds all round the coast and inland in the N. 43 cm.

AVOCET FAMILY   Recurvirostridae

Avocet *Recurvirostra avosetta*. Quite unmistakable at close quarters, with its black and white plumage and uniquely upcurved bill, but harder to pick out when in the water among a flock of gulls. One of the great successes of the RSPB (Royal Society for the Protection of Birds), which guards almost all its few British breeding places on the Suffolk coast. It was absent between the 1840s and 1940s. 43 cm.

PLOVER FAMILY   Charadriidae

Ringed Plover *Charadrius hiaticula*. Easily told from other shore birds by its black and white head and neck pattern and short, black-tipped, yellow bill. In flight a pale wing-bar shows. Breeds widely round the coast, on sand and shingle, and in some places inland. In winter also on muddy seashores and estuaries. Main call a liquid *too-i*. 19 cm.

Little Ringed Plover *Charadrius dubius*. A much scarcer, but not much smaller, bird of sand or shingle by fresh water, differing from the Ringed Plover especially by its lack of a wing-bar, mainly black bill and, at close range, yellow ring round eye. Mostly in NE England. Main calls *pee-oo* and *pip-pip*. 15 cm.

Golden Plover *Pluvialis apricaria*. A striking bird in summer with its golden brown upperparts and black and white underparts, but less distinctive in winter. Breeds on moorlands in the N and W, wintering in flocks on farmland and estuaries in the S, sometimes with Lapwings, when it is distinguished by its pointed wings. Call a piping *tlui*. 28 cm.

Grey Plover *Pluvialis squatarola*. A winter visitor only, to estuaries and mudflats, much less gregarious than the Golden Plover and differing especially in its black 'armpits' and very characteristic triple piping call, *tee-oo-ee*. In breeding plumage has silver grey, not golden brown, upperparts, but this is seen only in spring and early autumn. 28 cm.

164

Dotterel *Charadrius morinellus*. One of Britain's rarer breeding birds, confined to high mountain-tops in the Scottish Highlands, and a very few in the N of England and N Wales. Chestnut breast separates breeding adult from any other British breeding wader. On autumn passage to African winter quarters, white eye-stripe and breast-band should be looked for. 22 cm.

Turnstone *Arenaria interpres*. A common shore bird, best told by its short bill and legs and black and white appearance, also in summer by its rufous upperparts. Has never been proved to breed, but may be seen on the coast, and more rarely inland, at all seasons, especially on rocky shores. Turns over small stones and seaweed when searching for food. 23 cm.

Lapwing or **Peewit** *Vanellus vanellus*. The commonest inland breeding wader in lowland Britain, also common in hills throughout Britain and Ireland. Quite distinctive with its black and white plumage (the green only shows in good light), very rounded wings, and marked crest. 'Lapwing' refers to the lapping sound of its wings in display flight. The alternative name derives from its characteristic cry. 30 cm.

SNIPE FAMILY Scolopacidae
Dunlin *Calidris alpina* (1). Generally the commonest small shore bird, easily recognised in the breeding season by its black belly, but in winter could be confused with much whiter Sanderling. Breeds on wet moors and heaths and in coastal marshes. 17–19 cm. **Curlew Sandpiper** *C. ferruginea* (2) is slightly larger, with a more curved bill and a conspicuous white rump; uncommon passage migrant. 19 cm.

Little Stint *Calidris pusilla* (3). Another uncommon passage migrant, much smaller than the Dunlin and lacking its black belly. 13 cm. **Temminck's Stint** *C. temminckii* (4) is a still more uncommon passage migrant, but has bred in Scotland several times, and once in Yorkshire. More like a miniature Common Sandpiper, it differs from the Little Stint by its white outer tail feathers. 14 cm.

Knot *Calidris canutus*. A common winter visitor to the seashore, especially on sand or mud, feeding in large flocks, with the head characteristically held down. Intermediate in size between Dunlin and Redshank, it is most easily recognised in its reddish breeding plumage, sometimes seen in passage birds in May. Like many other winter visitor waders, it breeds in the Arctic. 25 cm.

Sanderling *Calidris alba.* A very characteristic bird of sandy seashores in winter, not often seen inland. It is an Arctic breeder, and its buffish breeding plumage is much less often seen than its bright white winter plumage. Will scurry along the tideline in front of human walkers. Dark shoulder-spot helps to distinguish it from Dunlin. 20 cm.

Purple Sandpiper *Calidris maritima.* A winter visitor to rocky shores which has only once been proved to breed here, although non-breeding birds may summer in the far N of Britain. Purple sheen of breeding plumage only visible at close range in good light. Yellow legs are safest field mark. Often accompanies Turnstones (p.165). 21 cm.

Redshank *Tringa totanus* (1). One of the commonest British shore birds, also not infrequent nesting inland; easily told by its broad, white wing-bars and red legs. Has a loud yelping cry. 28 cm. **Spotted Redshank** or **Dusky Redshank** *T. erythropus* (2) is a scarce passage migrant, with a longer bill, no white wing-bars, and an almost blackish breeding plumage. 30 cm.

Greenshank *Tringa nebularia.* A not uncommon shore bird on passage, also at fresh water inland, but breeds only on the moors of the Scottish Highlands, and rare on the coast in the S in winter. Green legs visible at close range, but best field marks are loud, piping *chu-chu-chu* call and large white patch on rump. 30–31 cm.

Common Sandpiper *Actitis hypoleucos.* Much the most likely small wader to be seen inland; a summer visitor, breeding widely by fresh water in the N and W, and frequent elsewhere on migration. Easily told by its typical flickering-wing flight low over the water and habit of bobbing its head up and down when at rest. Shows white wing-bar in flight. 20 cm.

Green Sandpiper *Tringa ochropus* (1). A not infrequent passage migrant by fresh water inland, occasionally seen in summer and winter also. Has nested. White rump conspicuous against dark upperparts in flight. 23 cm. **Wood Sandpiper** *T. glareola* (2) is less common on migration, but has bred more often, in Scotland. White rump less conspicuous and tail more barred. Call note, weaker than Green Sandpiper's, *weet-a-weet*. 20 cm.

Curlew *Numenius arquata*. Britain's commonest large grey–brown wader, with no special field marks apart from pale rump in flight; long, downcurved bill distinguishes it from godwits. Breeds commonly on moors, heaths and grassland, mainly in the N and W. Loud *cooorwee cooorwee* and *quee quee quee* notes are distinctive, as is its bubbling, musical song. 53–58 cm.

Whimbrel *Numenius phaeopus*. A smaller edition of the Curlew, from which it differs by its faster flight, tittering call and, at close range, 2 dark stripes and one pale one on head. Breeds only in northern and western isles of Scotland, and occasionally on nearby mainland, nesting on moorland. At other times on estuaries or sandy seashores. 41 cm.

Black-tailed Godwit *Limosa limosa*. Godwits are smaller than Curlews and have slightly upcurved bills. The Black-tailed is easily told by its black and white, almost Oystercatcher-like appearance in flight, and in summer by its reddish plumage. It now breeds in several parts of Britain, especially E Anglia, after disappearing in the 19th century. 41 cm.

Bar-tailed Godwit *Limosa lapponica*. Smaller than the Black-tailed and lacking its striking flight pattern, it is a winter visitor and passage migrant only, but at that time is commoner. Breeds in the Arctic, so its handsome reddish breeding plumage is rarely seen in Britain. Its almost straight bill distinguishes it from the Whimbrel in winter.

PHALAROPE FAMILY Phalaropodidae
Grey Phalarope *Phalaropus fulicarius*. Phalaropes are a distinct little group of waders which swim like miniature gulls. The Grey Phalarope is a scarce winter visitor, very rarely seen in its handsome red breeding plumage. It does not breed nearer than Iceland, and spends the winter on the open ocean. 20–21 cm.

Red-necked Phalarope *Phalaropus lobatus*. Britain's only breeding phalarope, a rare bird of the far northern and western extremities of the British Isles. Nests alongside shallow coastal lagoons. Easily told from Grey Phalarope in summer, but in winter its thin, needle-like bill is the best distinction. 18 cm.

167

Ruff *Philomachus pugnax*. Male a Redshank-like bird (but with variable bill and leg colour, usually yellowish or greenish) in winter plumage, as are females (Reeves) and immatures in summer. Males acquire striking and very variable plumage in breeding season, when they perform communal courtship displays. It ceased to breed in Britain last century, but is now recolonising, and has recently bred in 4 counties, mainly in E Anglia. Frequent on spring and autumn migration. 23–29 cm.

Ruffs and Reeves in flight

Woodcock *Scolopax rusticola*. One of the most widespread British breeding birds, nesting on the ground in woods and scrub, also a widespread winter visitor. Most often seen either when flushed from ground or 'roding' overhead in its curious slow territorial flight from March to July, uttering 2 distinct notes, one sibilant, the other almost growling. 34 cm.

Snipe *Gallinago gallinago*. Breeds in marshy areas throughout the British Isles, but less commonly in the S; also widespread and common by fresh water in winter. Recognised by its exceptionally long bill, and, when flushed, zigzag flight while uttering a harsh *creech creech*. In breeding season has remarkable 'drumming' or 'bleating' flight, diving at a steep angle. 27 cm.

Jack Snipe *Lymnocryptes minimus*. Markedly smaller than Snipe, from which it is also distinguished by a much shorter bill and habit, when flushed, of rising silently, flying more slowly and less erratically, and soon dropping down again. A winter visitor to fresh-water wetlands, seen much less often than Snipe. 19 cm.

THICK-KNEE FAMILY Burhinidae
Stone Curlew *Burhinus oedicnemus*. A striking bird, Curlew-like in distant flight and in wailing cry, but which differs from the Curlew in its short, straight bill, staring yellow eye, yellow legs and pale wing-bar. Decreasing, but still breeds in many places on chalk from Norfolk to Salisbury Plain, on the Suffolk coast and western South Downs. A summer visitor. 41 cm.

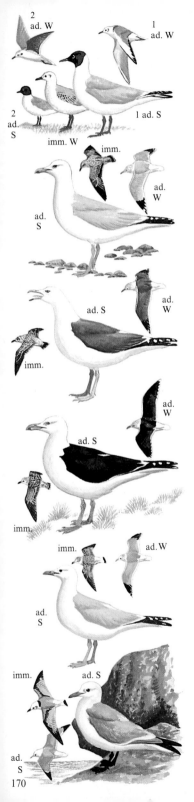

GULL FAMILY Laridae

Black-headed Gull *Larus ridibundus* (1). Generally the most frequently seen gull inland, and the smallest of the commoner gulls. The dark chocolate-brown hood of the breeding plumage, white blaze on fore-wing, and red bill and legs are distinctive. 35–38 cm. **Little Gull** *L. minutus* (2), rare but increasing in autumn and winter, is markedly smaller and has a dark underwing and an all-dark head in summer. 28 cm.

Herring Gull *Larus argentatus*. In general the commonest gull by the sea; also frequent inland. Combination of yellow bill and pink legs distinguish it from the smaller Common Gull. Back and wings much paler than Lesser Black-back, which has yellow legs. Breeds on sea cliffs, also sometimes on ground, but rarely inland. Nests on buildings in a few seaside towns. 56–66 cm.

Lesser Black-backed Gull *Larus fuscus*. Commoner inland than Great Black-back, but less so than Herring Gull; also widespread round the coast, breeding mainly on the ground and more often inland than Herring Gulls. Grey mantle varies from nearly as pale as Herring Gull to nearly as dark as Great Black-back, so combination of both bill and legs yellow is the safest field mark. 53–56 cm.

Great Black-backed Gull *Larus marinus*. The least common of Britain's 3 larger and commoner gulls, especially inland, where it very rarely breeds. It breeds mostly on cliffed coasts, and very occasionally between the Firth of Forth and the Isle of Wight. Large size, very dark blackish mantle and wings, and combination of yellow bill and pink legs are distinguishing features. 64–79 cm.

Common Gull *Larus canus*. A small edition of the Herring Gull, but with both bill and legs greenish yellow; it lacks the white fore-wing of the smaller Black-headed Gull. Widespread and common both inland and by the sea in winter, breeding inland and on the coast in Scotland and the NW part of Ireland; only rarely elsewhere. 41 cm.

Kittiwake *Rissa tridactyla*. Superficially like a Common Gull, but has all-black wingtips and black legs. Young birds have distinctive dark pattern on wings. Spends winter at sea, but breeds on sea cliffs, and exceptionally on buildings or ground by sea, all round coast. Rare inland. Loud chorus of *kitt-ee-waaake* gives rise to its name. 41 cm.

TERN FAMILY Sternidae

Sandwich Tern *Sterna sandvicensis*. The largest of our breeding terns, distinguished in flight at a distance by its size and harsh *kirrick* call, and at close range by its crest, black legs and yellow-tipped, black bill. Breeds on sand or shingle on most British and Irish coasts, except in Wales, SW England and Yorkshire. Like all British terns, a summer visitor and highly gregarious. 41 cm.

Common Tern *Sterna hirundo*. Generally Britain's commonest tern, breeding on sand or shingle on all coasts, except SW Wales and mainland SW England, also increasingly inland. Easily confused with Arctic Tern, but has black tip to bill in summer, slightly longer legs, and white breast and underparts contrasting more strongly with upperparts. 35 cm.

Arctic Tern *Sterna paradisaea*. Often commoner than Common Tern in the N and W, but much scarcer, except on migration, elsewhere; rarely breeds very far from the sea. Distinguished by its all-red bill, tinged purple rather than orange in summer, its shorter legs, and its much greyer underparts, nearer the colour of the upperparts. 35 cm.

Roseate Tern *Sterna dougallii*. Our least common breeding tern, with colonies mainly in Ireland and the Firth of Forth, but odd pairs often join Common and Arctic terneries. Its very long tail, extending well beyond the wing-tips, blackish bill with a patch of red at the base in summer, pinkish flush on breast (hard to see at a distance) and harsh *aach aach* call are the best field marks. 38 cm.

Little Tern *Sterna albifrons*. The smallest of our sea terns and the most endangered, since its breeding beaches, on all coasts except between Dorset and Cardigan Bay, are under increasing pressure from human beach users. Easily told by its small size, white forehead, and dainty, marionette-like flight, and at close range by its yellow legs and black-tipped, yellow bill. 24 cm.

Black Tern *Chlidonias niger*. Britain's only breeding marsh tern, which nests by fresh water, but very rarely and mainly in E England; not uncommon on spring and autumn migration. A very striking bird in breeding plumage, its black head and body contrasting with the grey wings; in winter resembles sea terns, but has a distinctive black spot on side of breast. 24 cm.

171

SKUA FAMILY   Stercorariidae

Great Skua *Stercorarius skua*. Skuas are like large, immature gulls with piratical habits, chasing other seabirds to make them drop their prey. The Great Skua is like an immature Great Black-back with a white wing-flash. Breeds only on N coast of Scotland, in northern isles and Outer Hebrides; seen on migration at sea further S. 58 cm.

Arctic Skua *Stercorarius parasiticus*. Smaller and more agile than the Great Skua, with projecting tail feathers when adult; 2 plumages, dark and light, the dark commoner in Britain. Breeds further S than Great Skua, to Jura, Inner Hebrides, and on inland moors in Caithness. Much more frequent offshore further S in autumn and winter. 46 cm.

AUK FAMILY   Alcidae

Razorbill *Alca torda*. Closest living relative of the extinct Great Auk; breeds commonly on cliffed coasts. In winter mainly at sea, except in SE. Its remarkable bill, flattened in a vertical plane, with a vertical white bar, immediately distinguishes it from the Guillemot. Like all auks, it has short wings, with a whirring flight, and dives constantly. 41 cm.

Guillemot *Uria aalge*. Generally the commonest auk, distinguished by its long, straight, dagger-like, all-dark bill, with plumage usually dark brown compared with Razorbill's black. Breeds on cliffs on all coasts except between Flamborough Head and the Isle of Wight. Most often seen in rafts on water off breeding cliffs; spends winter at sea. 42 cm.

Puffin *Fratercula arctica*. Breeds in burrows in huge colonies on cliffed coasts, except between Flamborough Head and Swanage. Decreasing because of pollution. Adult's most striking character is its extraordinary bill, which is less prominent in winter. Young birds have smaller bills, and can be told from young Razorbills by grey face. 30 cm.

Black Guillemot *Cepphus grylle* (1). A solitary breeder, unlike the gregarious Guillemot, commonest on cliffed coasts of Ireland, W Scotland and the northern isles, S to Banffshire in the E and Anglesey in the W. Breeding plumage very distinctive, all-black with large white wing-bars and red legs; much whiter in winter. 34 cm. **Little Auk** *Alle alle* (2) is an uncommon winter visitor, mainly to northern seas. 20 cm.

172

PIGEON FAMILY Columbidae

Rock Dove *Columba livia*. As a wild bird, the rarest British breeding dove, confined to wilder parts of the Scottish and Irish coasts. The numerous feral town pigeons are, however, their descendants, and many of these now breed on other parts of the coast. White rump, greyish wing-tips and broad black wing-bars distinguish it from the Stock Dove, much commoner in rural areas. 33 cm.

Stock Dove *Columba oenas*. Widespread and locally common in farmland, except in parts of Ireland and the Scottish Highlands, nesting in holes in old trees, buildings and both natural and artificial cliffs. Smaller than the Woodpigeon, and with no white, but 2 short black bars on each wing and blackish wing-tips. 33 cm.

Woodpigeon or **Ring Dove** *Columba palumbus*. Britain's largest pigeon or dove (the 2 names are interchangeable), breeding commonly throughout the British Isles in woods and farmland, and regarded as a serious agricultural pest. The white wing-bars and the adult's white neck-patch are diagnostic. Two white eggs are laid on a frail platform of twigs in a tree. 41 cm.

Collared Dove *Streptopelia decaocto*. A newcomer, which has only bred in Britain since the mid 1950s, but is now found everywhere from the Scillies to the Shetlands. Larger than the Turtle Dove, with duller grey plumage, much more white in the tail, and a black neck-bar. Song is persistent *coo-cooo-cuh*; compare the Woodpigeon's more soothing *coo-coo-coo, coo-coo*. 32 cm.

Turtle Dove *Streptopelia turtur*. Our only summer visitor dove, wintering in Africa; rare N of Yorkshire and in W Wales and Cornwall. Smaller than the Collared Dove, with more striking plumage, the upperparts chestnut, with a black and white patch on each side of the neck; tail more black than white. Voice a soothing *turr turrr*. 27 cm.

CUCKOO FAMILY Cuculidae

Cuckoo *Cuculus canorus*. Best known for its monotonous *cuck-oo* call from April to mid June, in flight it somewhat resembles a male Sparrowhawk, but its longer tail and more rakish outline help to distinguish it. Breeds throughout the British Isles, in all types of habitat, parasitising other birds, such as Hedgesparrow, Reed Warbler and Pied Wagtail, by laying its eggs in their nests. 33 cm.

ad. ♂

imm.
♀

173

OWL FAMILIES
Tytonidae and Strigidae

Barn Owl *Tyto alba*. The only white owl likely to be seen over most of Britain, the much larger, rare Snowy Owl being mostly a winter visitor to the far N. At rest appears mainly golden brown. Call is a long, blood-curdling screech; it also hisses, snores and barks. Nests in a hole, often in an old building. One of our most beneficial species, feeding on rats, mice and voles. 34 cm.

Long-eared Owl *Asio otus*. Britain's only owl with long ear-tufts, but these are not always visible; otherwise a slimmer version of the much commoner Tawny Owl, differing in its orange-yellow eyes. Breeds widely in woods and on moors and heaths, and is the only brown owl in Ireland. 36 cm.

Short-eared Owl *Asio flammeus*. The most likely larger owl to be seen flying by day, when its dark wing-patch makes it look like a giant eyed moth. Ear-tufts are very short and hard to see; yellow eyes separate it from the Tawny Owl. A locally frequent breeder in Scotland, N England, mid Wales, and along the E coast S to the Thames estuary; more widespread in winter. 38 cm.

Tawny Owl *Strix aluco*. Generally Britain's commonest owl, widespread in woodland in England, Scotland and Wales, extending into well-treed towns, villages and suburbs. Black eyes distinguish it from the 2 eared owls; call is a long-drawn-out hoot, also a sharp *ke-wick*, *ke-wick*. Not often seen by day, but can be found at roost by following up scolding Jays and Blackbirds. 38 cm.

Little Owl *Athene noctua*. The smallest British breeding owl, not a native, but introduced from Europe about 100 years ago; now widespread in England, less so in Wales, and just beginning to spread into Scotland. Often seen by day, it has a curious habit of bobbing and waggling its head from side to side. Call is a loud, ringing *kiew kiew*. 22 cm.

NIGHTJAR FAMILY   Caprimulgidae
Nightjar *Caprimulgus europaeus*. A decreasing species, because of the diminution of its heathland habitat, now mainly confined to England S of the Thames. Rarely seen by day, except when flushed from its nest or roost on the ground; usually detected by its penetrating churring call, not usually heard until at least three-quarters of an hour after sunset. When glimpsed at dusk it resembles a giant Swift. 27 cm.

174

SWIFT FAMILY Apodidae
Swift *Apus apus*. A common summer migrant, breeding in crevices in buildings throughout the British Isles apart from the far N of Scotland. Long wings, short tail, all-dark plumage and habit of screaming in parties on summer evenings all distinguish it from the Swallow and martins. Only present in any numbers from mid May to early August (Swallow mid April to early October). 16–17 cm.

KINGFISHER FAMILY Alcedinidae
Kingfisher *Alcedo atthis*. One of the handsomest British birds, most often seen as a sapphire flash as it speeds down a river or across a pond. Widespread, but uncommon in Scotland, breeding in holes by or near fresh water of all kinds; frequents the seashore in winter. Subject to big population fluctuations due to hard winters, which freeze up its feeding places. 16–17 cm.

WOODPECKER FAMILY Picidae
Green Woodpecker *Picus viridis*. The most conspicuous of our 4 woodpeckers, its yellow rump showing up as it flies away from the observer, uttering its characteristic yelping cry. A common breeding bird in England and Wales, usually in holes in trees, but thins out in Scotland, not extending N of the southern Highlands; not in Ireland. Often feeds on ants on the ground. 32 cm.

Great Spotted Woodpecker *Dendrocopus major*. The larger of our 2 black and white woodpeckers, easily told by its size, red vent and red nape-patch (absent in the female). Has a curious rattling 'song', made by pecking vigorously at resonant dry wood. Breeds in woodland and among scattered trees throughout Great Britain except the far N; not in Ireland. 23 cm.

Lesser Spotted Woodpecker *Dendrocopus minor*. Much smaller than the Great Spotted (sparrow-sized), and further distinguished by barred back and male's red crown. Also drums, but more weakly. Confined to England and Wales, becoming rare N of Yorkshire and the Lake District; much less often seen than the Great Spotted; commonest to the S and W of London. 14–15 cm.

Wryneck *Jynx torquilla*. Once a common summer visitor to S and Midland England, our only brown woodpecker is now one of our rarest breeding birds, and almost extinct in the S. Recently it has begun to colonise Scotland, perhaps from Norway. A small, rather long-tailed bird, little bigger than a Sparrow, it has a curious way of twisting its neck round. 16–17 cm.

175

LARK FAMILY Alaudidae

Shore Lark *Eremophila alpestris*. An uncommon winter visitor, mainly to the E coast. The Shore Lark's yellow face pattern is much less obvious in winter than in summer, when it is rarely seen. The call is a *tseep*, like that of the Rock Pipit or Yellow Wagtail. 16–17 cm.

Woodlark *Lullula arborea*. One of Britain's finest songsters. Its mellow, fluty song is quite distinct from the Skylark's, but also delivered on the wing. The Woodlark also differs in its shorter tail and habit of perching in trees. Decreasing on heaths and in other country with scattered trees; now mainly found S of the Thames, with outliers in E Anglia and S Wales. 15 cm.

Skylark *Alauda arvensis*. One of Britain's most famous song-birds, thanks to numerous poets. Its song is characteristically delivered while the bird hangs high in the air. It has a rather inconspicuous crest, but otherwise is a classic 'little brown bird'; young birds have short tails, like Woodlarks. Widespread and common in all kinds of open country, farmland, moors, heaths and dunes. 18 cm.

SWALLOW FAMILY Hirundinidae

Swallow *Hirundo rustica*. One of the commonest and most familiar of British wild birds, this summer visitor is identified by its long, forked tail (shorter in young birds), all blue–black upperparts, and chestnut throat and forehead. Breeds on ledges inside buildings throughout the British Isles. Arrives in early April. 19 cm.

Sand Martin *Riparia riparia*. Plumage is quite different from both Swallow and House Martin, being sandy brown above, and white with a brown chest-bar below; tail short. Nests in tunnels excavated in sandy or gravelly banks of rivers or man-made pits, throughout Britain and Ireland except Outer Hebrides and northern isles. Arrives before the Swallow, in late March. 12 cm.

House Martin *Delichon urbica*. Easily distinguished by its white rump, short tail and all-white underparts. Builds a mud nest on the outside of buildings or in caves. A summer visitor to the whole of Britain and Ireland, thinning out in the far N of Scotland, arriving a little later than the Swallow, in mid or late April. 12–13 cm.

WAGTAIL FAMILY Motacillidae

Tree Pipit *Anthus trivialis*. A summer visitor that breeds almost throughout the mainland of Great Britain but not at all in Ireland, nesting on the ground in open woodland and other places with scattered trees, which it needs for a take-off point for its brief, trilling song, which ends in *see-er see-er see-er*. Very like a Meadow Pipit, but usually has darker legs. 15 cm.

Meadow Pipit *Anthus pratensis*. A very common resident of moors, heaths and open country throughout Britain and Ireland. Resembles a diminutive Thrush, and differs from Tree Pipit in not using a tree for its song flight, having a shriller call-note, a song without the Tree Pipit's distinctive ending, flesh-pink legs and a much longer hind claw (hardly detectable in the field). 14–15 cm.

Rock Pipit *Anthus spinoletta*. A greyer bird than either the Tree or Meadow Pipit, and distinguished from both by its greyish, not white, outer tail feathers, and by generally occupying a coastal habitat, especially on rocky shores. Call a single *pheet*, unlike the Meadow Pipit's usual triple call. Absent only between Flamborough Head and North Foreland, and even there frequent in winter. 16–17 cm.

Pied Wagtail *Motacilla alba*. Wagtails are well known for their habit of wagging their tails up and down, and for their bounding flight. This is the only black and white species in Britain and Ireland, widespread and common at all seasons. Often in or near towns, suburbs and villages, sometimes nesting in holes and crannies in buildings. 18 cm.

Grey Wagtail *Motacilla cinerea*. A bird of swift mountain streams, also found by many lowland ones, except in much of E England; mainly a summer visitor to the N and W. Breeding male very distinctive with its black gorget, and at all times greyer on the back than the Yellow Wagtail; it also has a longer tail and a louder, more metallic call-note. 18 cm.

Yellow Wagtail *Motacilla flava* (1). A summer visitor to England, except for most of the SW, the E and parts of S Wales and S Scotland. Usually breeds in marshes or by fresh water, but sometimes in drier places. Breeding male is most brilliantly yellow of all British birds. Call-note, *tsweep*, is softer than other Wagtails'. 16–17 cm. **Blue-headed Wagtail** (2) is Continental race, sometimes seen on passage.

177

WAXWING FAMILY Bombycillidae

Waxwing *Bombycilla garrulus*. Irregular winter visitor, sometimes visiting berried trees and bushes in suburban gardens and town parks. Crest conspicuous, waxy red wing-spots and yellow tail-tip less so, except at close range. Flight not unlike Starling's. Used to be called Bohemian Chatterer, after one of its call-notes. 18 cm.

SHRIKE FAMILY Laniidae

Red-backed Shrike *Lanius collurio* (1). Increasingly scarce summer visitor, now largely confined to scrubby areas in SE England, and especially E Anglia. Also called Butcher Bird from its habit of impaling insects, nestlings and other prey on thorns. 17 cm. **Great Grey Shrike** *L. excubitor* (2) is a scarce winter visitor, usually seen perched high in a tall shrub. 24 cm.

STARLING FAMILY Sturnidae

Starling *Sturnus vulgaris*. A widespread and very common bird, breeding throughout Britain and Ireland, often in towns, where winter visitors may roost in trees or on buildings. Distinguished from cock Blackbird by iridescent (summer) or spotted (winter) plumage, and bustling gait, walking, not hopping. Juveniles are ash-grey. 21–22 cm.

ACCENTOR FAMILY Prunellidae

Hedgesparrow or **Dunnock** *Prunella modularis*. A common breeding bird throughout the British Isles, except Shetland, in woodland, scrub, farmland, and often in gardens and town parks. Thin bill, grey breast and speckled flanks all distinguish it from the House Sparrow. Call is a high-pitched *tseep*; song a rather flat little warble. 14–15 cm.

WREN FAMILY Troglodytidae

Wren *Troglodytes troglodytes*. One of Britain's smallest birds, widespread and common in a wide range of habitats, from town parks to mountainsides, throughout the British Isles. It can readily be distinguished from other small brown birds by its warm brown plumage and habit of cocking up its remarkably short tail. Song is a loud and vigorous burst of clear, trilling notes. 9–10 cm.

DIPPER FAMILY Cinclidae

Dipper *Cinclus cinclus*. One of the most distinctive British and Irish birds in both plumage and habitats. Frequents fast streams, mainly in the hill districts of the N and W, but also as far SW as the Cotswolds, often building its domed nest under a waterfall. Its white breast, short tail and habit of bobbing up and down on a stone in midstream are diagnostic characters. 18 cm.

WARBLER FAMILY Sylviidae

Grasshopper Warbler *Locustella naevia.* One of the birds that is much more often heard than seen. Its high-pitched grasshopper-like trill (actually more like an angler's reel) can often be heard in dense vegetation when the bird itself refuses to emerge. When it does, look for the speckled back and graduated tail. A summer visitor, like almost all our warblers, breeding throughout Britain and Ireland. 13 cm.

Savi's Warbler *Locustella luscinioides.* Used to inhabit the undrained fens, but became extinct in 1856; since 1960 it has begun to breed again in Kent and E Anglia. Larger and less skulking than the Grasshopper Warbler, it has unstreaked upperparts, a white chin, and a slower, less high-pitched song, uttered in shorter bursts. 14 cm.

Reed Warbler *Acrocephalus scirpaceus.* A fairly common breeding bird in reed-beds in S and central England, also in bushes and tall vegetation. Most often detected by its churring, mimetic song, like 2 pebbles being rubbed together. Extremely like the Marsh Warbler, but has slightly more rufous plumage, duller throat and darker legs. 12–13 cm.

Marsh Warbler *Acrocephalus palustris.* A rare breeder of a few southern localities, mainly in the Severn valley, so like the Reed Warbler that it is best told by its more musical and still more mimetic song, mimicking many other birds. It also has a whiter throat, more olive-brown upperparts and pinker legs. Frequents a variety of bushy places, often singing from a prominent perch. 12–13 cm.

Sedge Warbler *Acrocephalus schoenobaenus.* Easily told from Reed and Marsh Warblers by its conspicuous white eye-stripe, and by its varied song, which has more harsh notes than the other 2, and is also mimetic. Breeds in bushy places, usually near water, but sometimes in tall vegetation, such as willow-herb or nettles, almost throughout Britain and Ireland. 13 cm.

Cetti's Warbler *Cettia cetti.* A very recent colonist of S England from the nearby Continent, most often detected by its loud song, uttered in staccato bursts, *chewee chewee chewee weeweewee.* A skulker, when glimpsed it has dark brown plumage with a rounded tail. Usually in thick vegetation by fresh water. One of Britain's 2 resident warblers. 14 cm.

179

Willow Warbler *Phylloscopus trochilus*. One of our commonest breeding birds, found throughout Britain and Ireland almost wherever there are trees or shrubs. Very like the Chiffchaff, but has slightly yellower underparts, especially when immature, and usually paler, flesh-coloured legs; the best distinctions are its mellifluous song and its *hoo-eet* call. 11 cm.

Chiffchaff *Phylloscopus collybita*. Almost as common as the Willow Warbler in the S, but much less so in Scotland, and with similar habitat needs. Best distinctions are its monotonous *chiff-chaff chiff-chaff* song and monosyllabic call-note *hweet*, but it also appears browner, with less green and yellow in its plumage, and always has dark legs. One of the earliest migrants to arrive, in late March. 11 cm.

Wood Warbler *Phylloscopus sibilatrix*. The largest and yellowest of Britain's 3 leaf warblers, also distinctive for its white underparts and especially for its 2 songs, a long, quivering trill and a plaintive, bell-like note. Breeds in woods throughout Great Britain, but especially in the W; almost absent from Ireland. 12–13 cm.

Goldcrest *Regulus regulus*. The smallest British and Irish bird, like a miniature leaf warbler, with a tiny, needle-like bill, the male having an orange and the female a yellow crest. Breeds almost throughout Britain and Ireland, in woods, scrub, gardens and parks, wherever there are a few pines, spruces, yews or other conifers. 9 cm.

Firecrest *Regulus ignicapillus*. A recent colonist from the Continent that now breeds in a few scattered woodlands in SE England. Almost as tiny as the Goldcrest, it can be told by its distinctive white eye-stripe, as well as by its less high-pitched call-note and song. Also an uncommon winter visitor. 9 cm.

Dartford Warbler *Sylvia undata*. Until the arrival of Cetti's Warbler in the 1970s, this was the only resident British warbler. It has always been confined to heathland with gorse and heather S of the Thames, and suffers greatly in hard winters; now largely confined to Dorset and the New Forest. A shy, skulking bird, it is most often seen in its dancing display flight. 12–13 cm.

180

Whitethroat *Sylvia communis.* A widespread breeding bird of heaths and scrubland, including hedgerows, now much less common than it was in the 1960s, due, it is thought, to drought in its African winter quarters. White throat, rufous wings, male's grey head, and dancing song flight are its best field marks. 14 cm.

Lesser Whitethroat *Sylvia curruca.* As common as Whitethroat in some southern districts, but thinning out northwards and westwards and absent from Scotland and Ireland. Prefers taller scrub to the Whitethroat, from which it differs in its greyer plumage, dark cheeks and especially its tuneless little rattling song. 13–14 cm.

Garden Warbler *Sylvia borin.* The archetypal small brown bird, having virtually no distinguishing plumage feature, and best recognised by its song, an even, musical warble, lower-pitched and more continuous than the Blackcap's. Breeds in woods and scrub throughout Great Britain, but rare in the Scottish Highlands; a few pairs also breed in Ireland. 14 cm.

Blackcap *Sylvia atricapilla.* Male's black and female's brown crown distinguish it from all other warblers; also larger and longer-billed than Marsh and Willow Tits. Song a musical, high-pitched, rather staccato warble. Distribution in Great Britain similar to Garden Warbler, but more frequent in Ireland. Shows increasing tendency to winter, when most often seen at bird tables. 14 cm.

FLYCATCHER FAMILY Muscicapidae
Spotted Flycatcher *Muscicapa striata.* Britain's commonest flycatcher, easily told by its streaked breast (it is the juveniles that are spotted) and habit of flying out from a perch to catch flying insects and back again. A summer visitor, breeding throughout Britain and Ireland, except the Outer Hebrides and northern isles. 14 cm.

Pied Flycatcher *Ficedula hypoleuca.* Male is strikingly black and white, but female and young birds much more like the larger Spotted Flycatcher. A summer visitor, breeding especially in oakwoods in Wales, SW and NW England and S Scotland; rather uncommon in the Highlands. Readily breeds in nest boxes in these areas. 13 cm.

181

THRUSH FAMILY Turdidae

Stonechat *Saxicola torquata*. A resident bird of heaths, moors and coastal cliffs, common all round the coast, though less so in E England, and inland in Ireland, W Scotland and S England. Male easily told by its distinctive black, white and chestnut plumage pattern; has dancing song flight. Female has darker throat and tail than female Whinchat. 12–13 cm.

Whinchat *Saxicola rubetra*. A summer visitor, more often on farmland and inland than Stonechat, and much less common in Ireland and along the S coast of England. Male is easily told by white eye-stripe and much less black on head than Stonechat; female has pale throat and white sides to tail. Both chats show white wing-bar in flight. 12–13 cm.

Wheatear *Oenanthe oenanthe*. A common summer visitor to moors, heaths, dunes and other rough ground in the N and W, nesting in holes in the ground; local in SE England. White rump distinguishes both sexes as they fly away. Male has black, white and grey head pattern, especially handsome in spring. One of the earliest migrants to arrive, in late March. 14–15 cm.

Redstart *Phoenicurus phoenicurus* (1). A widespread summer visitor to Great Britain, but local in SE England and rare in Ireland; frequents woods, heaths and parkland. 14 cm. **Black Redstart** *P. ochrurus* (2) is a scarce breeder in SE England and a more widespread passage migrant, frequenting built-up areas and sea cliffs. 14 cm. Both are readily told by their red tails.

Robin *Erithacus rubecula*. One of Britain's most familiar and best-loved birds, and the only one with the whole face as well as the breast orange–red. Young birds with their speckled breast look very different, like miniature Thrushes. Breeds in woods, scrub, parks, gardens and farmland throughout Britain and Ireland, except Shetland. Sings almost throughout the year. 14 cm.

Nightingale *Luscinia megarhynchos*. The most famous British songster, which sings almost as much by day as by night. A summer visitor to the SE half of England, breeding in dense scrub in woodland and on heaths and commons. Rarely seen, except sometimes for its rufous tail diving into the nearest bush. Nests almost on the ground. 16–17 cm.

Blackbird *Turdus merula*. A well-loved song bird; the cock is very distinctive, lacking the spots or sheen and bustling gait of the Starling. The female is more thrush-like but has only indistinct spots on the breast. Breeds in woods, scrub, farmland, parks and gardens throughout Britain and Ireland, from high moors to town centres. Nest lined with dried grass; eggs blue–green with brown markings. 25 cm.

Ring Ouzel *Turdus torquatus*. The upland Blackbird – though Blackbirds are found on the moors too. A summer visitor to hill districts in the N and W, but rather rare in Ireland. Easily told by the white gorget, much more conspicuous in the male than in the female; also has pale patch on wing. Arrives in late March. 24 cm.

Fieldfare *Turdus pilaris*. The larger of our 2 winter visitor thrushes, told by its blue–grey head and rump contrasting with the chestnut back; also shares the Mistle Thrush's white underwing flashing in flight. Widespread and common throughout Britain and Ireland, especially on farmland; usually in towns only in hard weather. Flight-note a chuckling *chack chack*. 25–26 cm.

Redwing *Turdus iliacus*. The smaller of Britain's 2 winter visitor thrushes, told from the Song Thrush by its white eye-stripe and red flanks and underwing. Widespread and common in farmland, often associating with Fieldfares. Also breeds rather locally in parts of the Highlands of Scotland. Flight-note a high-pitched *seeih*. 21 cm.

Song Thrush *Turdus philomelos*. One of the commonest and most widespread garden birds in Britain and Ireland, also in woods, farmland and other open country with trees and bushes. Best recognised by its size (smaller than Blackbird), conspicuously spotted breast, and loud, clear, repetitive song. Nest is lined with mud; eggs are blue with black spots. 23 cm.

Mistle Thrush *Turdus viscivorus*. The largest British and Irish thrush, greyer than the Song Thrush and with white tips to outer tail feathers, bolder breast-spots, white underwing flashing in flight, and more ringing, Blackbird-like song. Widespread and common in farmland, parkland and large gardens. Nest is lined with dried grass; eggs are white with brown spots. 27 cm.

183

TIT FAMILY Paridae

Coal Tit *Parus ater* (1). The smallest British and Irish tit, easily identified by the prominent white patch on its nape, also by its high-pitched *ticha ticha* call, like a miniature of the Great Tit's well-known *teacher teacher*. Widespread and common, breeding everywhere except the Outer Hebrides and northern isles, wherever there are a few pines or other conifers. 11–12 cm.

Great Tit *Parus major* (2). The largest British and Irish tit, sparrow-sized, readily recognised by its white cheeks on a black head and broad, black bib down a yellow breast. Widespread and common wherever there are trees, apart from the Outer Hebrides and northern isles (which have few trees). Best-known call a loud *teacher teacher*, also one like a saw being sharpened. 14 cm.

Blue Tit *Parus caeruleus* (3). The only British and Irish bird with a substantial amount of blue in its plumage. Generally the commonest tit, with the same distribution as the Great Tit. Most familiar call a churring *tsee tsee tsee tsit*. Like most tits it nests in a hole in a tree or building and readily takes to nest boxes. 11–12 cm.

Marsh Tit *Parus palustris*. The commoner of Britain's 2 black-capped tits, easily told from the Coal Tit by lack of white nape-patch, but much harder to separate from the Willow Tit. Has characteristic *pitchüü* and *chicka bee bee bee* calls. A common woodland bird in England and Wales, but only extending into extreme SE Scotland; absent from Ireland. 11–12 cm.

Willow Tit *Parus montanus*. Less common but more widespread than the Marsh Tit, extending N to the Forth–Clyde line and occasional in the Highlands. Distinguished from the Marsh Tit by a faint, pale patch on the wing and slightly less black cap, but more effectively by its harsh, nasal *tchay* or *aig* call. 11–12 cm.

Long-tailed Tit *Aegithalos caudatus*. A very distinctive small tit, usually seen in a party, often flying one after another from bush to bush, easily recognised by their long tails. Widespread and fairly common throughout Britain and Ireland, except the Outer Hebrides and northern isles. Has a most distinctive oval nest, beautifully woven from moss and cobwebs. 14 cm.

Crested Tit *Parus cristatus*. Britain's rarest tit, confined to pinewoods and coniferous plantations in the Scottish Highlands, especially along Speyside. Easily recognised by its crest; indeed it is the only small British bird to have a crest. Call a soft trill and the usual tit contact note, *si-si-si*. 11–12 cm.

BABBLER FAMILY   Timaliidae
Bearded Tit *Panurus biarmicus*. Once thought to be a tit, but now known to be allied to the babblers of tropical Africa and Asia, this is one of Britain's most local breeding birds, confined to extensive reed-beds, mainly along the coast, from the Humber to Dorset. The long tail distinguishes it from other small reed-bed birds, and so does the male's moustache. 16–17 cm.

NUTHATCH FAMILY   Sittidae
Nuthatch *Sitta europaea*. Easily recognised by its blue–grey and deep buff plumage, as well as by its habit of running up and down tree-trunks; no other bird habitually descends trees head-downwards. Frequent in England and Wales, but thinning out northwards and absent from Scotland and Ireland. Wedges nuts into crevices and hammers them open. 14 cm.

CREEPER FAMILY   Certhiidae
Treecreeper *Certhia familiaris*. The only small land bird in Britain and Ireland with a curved beak, and certainly the only one that climbs trees, indeed being hardly ever seen away from tree-trunks. Widespread and frequent in woods and other well-treed places, including large gardens. Nests in remarkably narrow crevices, often between ivy stems and a tree-trunk. 12–13 cm.

BUNTING FAMILY Emberizidae
Corn Bunting *Miliaria calandra*. The largest British bunting, and one of the least distinguished-looking British birds with its plain brown plumage relieved only by dark specklings. A bird of open, often arable, country, widespread and locally frequent, but absent from large areas of Wales, Ireland, the Highlands and SW England. Has a very distinctive, high-pitched, jangling song. 18 cm.

Yellowhammer *Emberiza citrinella*. Generally the commonest bright yellow bird of British and Irish farmland, well known for its monotonous, high-pitched song, usually rendered as *a little bit of bread and no cheese*. Breeds throughout Britain and Ireland, except for Shetland. Females are browner and less yellow. 16–17 cm.

Cirl Bunting *Emberiza cirlus*. A very local-breeding bird, hardly found nowadays except S of the Thames, and most frequent not far from the sea along the S coast. Male is like a Yellowhammer with black on its head; female hardly distinguishable unless olive instead of chestnut rump can be seen. Song is quite different, more like the Lesser Whitethroat's tuneless rattle. 16–17 cm.

Reed Bunting *Emberiza schoeniclus*. The bunting of the wetlands, most often found in reed-beds or by fresh water, but increasingly also in quite dry farmland. Cock is unmistakable with its striking black head-dress; hen is sparrow-like but distinguishable by white outer tail feathers, readily seen in flight. Squeaky staccato song also easy to recognise. 15 cm.

Snow Bunting *Plectrophenax nivalis*. An extremely rare breeding bird of high mountain-tops in Scotland, much commoner as a winter visitor, mainly to the E coast, but also to hill-tops in the N. A flock of Snow Buntings is aptly named 'snowflakes', as their black and white wings flash on and off in flight. Females and winter males have some brown in their plumage. 16–17 cm.

FINCH FAMILY Fringillidae
Hawfinch *Coccothraustes coccothraustes*. Our largest finch has the most massive bill of any British bird for its size, and uses it to crack fruit stones. A widely scattered breeding bird, commonest in the SE, it extends N to the fringe of the Highlands, but is absent from Ireland. The Robin-like *tick* note, once learned, is a useful means of detecting its dumpy form flying overhead. 18 cm.

186

Brambling *Fringilla montifringilla*. A winter visitor, most often seen under Beech trees, on whose nuts it likes to feed, but sometimes also on farmland with other finches; has bred once, in Scotland in 1920. Recognisable at all times by its white rump as it flies away, also by orange–buff breast and shoulders. Usual call a harsh *tsweek*. 14–15 cm.

Chaffinch *Fringilla coelebs*. Together with the Blackbird, one of the 2 commonest British birds, breeding in woods, scrub, hedgerows, parks and gardens throughout Britain and Ireland and, at least locally, on most of the islands. White shoulder-patch is a good field mark, and male's handsome plumage is distinctive. Usual call, *pink pink*, is confusable with Great Tit's, but cheery song is not. 15 cm.

Goldfinch *Carduelis carduelis*. Widespread and frequent throughout Britain and Ireland, but uncommon in N half of Scotland and absent from northern and western isles. Has unique combination of black, white, red and yellow plumage, also showing white in rump in flight. Well known for its habit of feeding on thistles. Has a conspicuously dancing flight and a liquid call-note. 12 cm.

Siskin *Carduelis spinus*. A small, greenish finch that breeds throughout the Scottish Highlands and much more locally in the rest of Britain and Ireland, especially in the S. Yellow rump of both sexes and black crown and chin of male are good field marks. Much commoner as a winter visitor, especially visiting Alders with Redpolls. 12 cm.

Greenfinch *Carduelis chloris*. One of our commonest and most widespread garden and village birds, also flocking in winter on farmland with other finches. Breeds throughout Britain and Ireland, but more local in N Scotland. Bright yellow wing-patches are good field mark, together with yellow rump and sides to tail, set in a generally dull greenish yellow plumage. 14–15 cm.

Bullfinch *Pyrrhula pyrrhula*. A very distinctive bird with its black crown, rosy breast and cheeks and white rump as it flies away; females are similar but duller. Widespread and frequent in woods, orchards, parks and gardens throughout Britain and Ireland, but less so in far N of Scotland. Disliked by orchard owners because it eats fruit buds. 14–16 cm.

Lesser Redpoll *Carduelis flammea*. A small brown finch with a red forehead, distinguished from both Linnet and Twite by its black chin. Widespread and frequent throughout Britain and Ireland, especially in winter, when it feeds in riverside Alders with Siskins; also in breeding season, but then still rather local in parts of the S. Usual flight-note *chuch-uch-uch*. 13–15 cm.

Twite *Carduelis flavirostris*. The moorland Linnet, distinguished from both Redpoll and Linnet by its pink rump and lack of red crown; shows a whitish wing-bar in flight. Most frequent in N and W Scottish Highlands, also along W coast of Ireland and on the Pennine moorlands of N England; elsewhere a winter visitor, mainly to the coast. 13–14 cm.

Linnet *Carduelis cannabina*. The commonest and most widespread of small brown British and Irish finches, but local and uncommon in parts of the Highlands and islands. Differs from Redpoll and Twite by its grey head and chestnut back. A bird of heaths and scrub, but found also in a wide variety of man-made habitats, often breeding in small colonies like the Greenfinch. 13–14 cm.

Crossbill *Loxia curvirostra*. Unique among British birds for its crossed mandibles, though these are only visible at close range. Cock is the only all-red small song-bird; young males are orange–red and females greenish yellow. Breeds regularly among pines in Scottish Highlands, elsewhere also among other conifers after periodic irruptions from the Continent. 16–17 cm.

SPARROW FAMILY Passeridae
Tree Sparrow *Passer montanus*. Distinguished from the House Sparrow by both sexes having an all-chestnut crown and a black cheek-patch. Nests in small colonies in holes in trees, cliffs, quarries and buildings, not usually in towns or villages. Widespread and frequent, but thinning out northwards and westwards and only near the coast in Ireland. 14 cm.

House Sparrow *Passer domesticus*. Only the cock bird has the grey crown and black bib, hens being the archetypal small brown bird, easily recognised by their cheerful chirrup. Widespread and common throughout Britain and Ireland, including all kinds of built-up areas, but rather local in under-populated parts of the Scottish Highlands. Usually nests in a hole, but occasionally in a tree or bush. 14–15 cm.

CROW FAMILY Corvidae

Jay *Garrulus glandarius*. One of Britain's most brightly coloured birds, it looks almost tropical as it slips into the trees with a harsh cry, revealing its white rump. Common in woodlands in England and Wales, but much less so in Scotland and Ireland, and absent from most of the Highlands. In autumn flies to and from Oak trees, collecting acorns to bury. 34 cm.

Magpie *Pica pica*. Unmistakable with its long tail and black and white plumage. Chief call a chattering chuckle. Widespread and common in farming districts, and sometimes in villages and suburbs, in England, Wales and Ireland; in Scotland largely absent from the mountain and moorland districts of the Highlands and the southern uplands. 46 cm.

Jackdaw *Corvus monedula* (1). A common and gregarious small crow of farmland and sea cliffs, easily told by its grey nape and *chack chack* call. Found almost throughout Britain and Ireland, but local in NW Scotland. 33 cm. **Chough** *Pyrrhocorax pyrrhocorax* (2). Confined to sea cliffs in Wales, Ireland, Islay and the Isle of Man; rarely inland; extinct in Cornwall. Easily told by its red bill and legs. 39–40 cm.

Raven *Corvus corax*. The largest British crow, all-black, and best told by its size, heavy bill and hoarse croak. Frequent in Wales, SW and NW England, most of Scotland, except the E coast, and most of Ireland except the centre. Inhabits moors, mountains and sea cliffs, building its substantial nest on a ledge, the eggs being laid as early as February in some years. A fine aerial acrobat. 64 cm.

Carrion Crow *Corvus corone corone* (3) and **Hooded Crow** *C. c. cornix* (4). Two races of a common and widespread bird, covering the whole of Britain and Ireland, but with a dividing line that leaves the Hooded Crow in Ireland, the Isle of Man and W Scotland. Grey back makes the Hooded Crow unmistakable; the all-black Carrion Crow differs from the Rook in its feathered face and harsher voice. Both 47 cm.

Rook *Corvus frugilegus*. Our most colonial crow, rookeries being a common feature of lowland agricultural Britain and Ireland. Rooks are always more gregarious than crows (which can flock) and have a conspicuous bare patch at the base of the bill and longer feathers on the thigh, giving a 'baggy trousers' appearance. Often seen with Jackdaws. 46 cm.

189

SOME COLOURFUL VISITORS

Bee-eater *Merops apiaster* (1). An occasional wanderer from the Mediterranean, it bred in Sussex in 1955. 28 cm.

Bluethroat *Luscinia svecica* (2). A Robin-like bird that attempted to nest in Scotland in 1968. A not infrequent passage migrant on the E coast. 14 cm.

Roller *Coracias garrulus* (3). A large, bright, crow-like bird that occasionally wanders to Britain from S or E Europe. 31 cm.

Golden Oriole *Oriolus oriolus* (4). A bird that visits Britain regularly in small numbers, and occasionally breeds in a few scattered localities in the S. 24 cm.

Nutcracker *Nucifraga caryocatactes* (5). A speckled brown crow that irrupts at irregular intervals from E Europe. 32 cm.

Ring-necked Parakeet *Psittacula krameri* (6). An escape from captivity that is already breeding in and S of the Thames valley in tree holes; it frequents mainly parks and gardens. 41 cm.

Hoopoe *Upupa epops* (7). A very rare breeder in S England, usually in a tree hole, not every year. More frequent as an occasional visitor in spring and autumn. 28 cm.

190

Other Vertebrates

Below the mammals and birds in the evolutionary tree there are 3 other main groupings of vertebrates: the reptiles, amphibians and fish. All are popularly called 'cold-blooded', though in fact their temperature varies according to that of their surroundings, instead of being stoked to a constant level from within, like that of mammals and birds.

Common Toad, and Common and Crested Newts are widespread and common (though the Frog is believed to be only introduced in Ireland); the Natterjack Toad and Palmate Newt are much more local.

Fishes are gill-breathing vertebrates, comprising 4 main classes, each of equal evolutionary status to the reptiles and amphibians: Marsipobranchii (lampreys, hagfishes), Selachii (sharks, dogfish), Bradyodonti (chimaeras), and

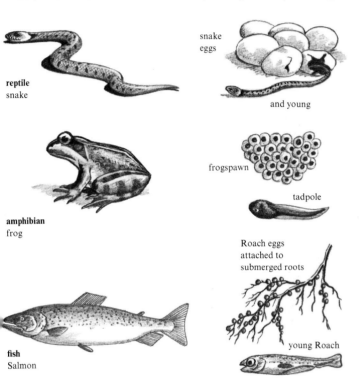

reptile
snake

snake eggs

and young

amphibian
frog

frogspawn

tadpole

Roach eggs attached to submerged roots

fish
Salmon

young Roach

Reptiles are a class of horny-skinned vertebrates, very poorly represented in Britain by 3 snakes – the Grass Snake, the very local Smooth Snake, and the Adder or Viper – and 3 lizards – the Common Lizard (which is the only reptile that extends to Ireland), the very local Sand Lizard, and the Slow-worm, which looks like a snake and is often thought by the uninitiated to be one, but is in fact a legless lizard.

Amphibians differ from reptiles in their softer skin and their need to resort to water to breed, because the young pass through a gill-breathing tadpole stage. There are 6 native British species – 3 newts, 2 toads and a frog. The Frog,

Pisces (the true or bony fishes). It is to this last group that the great bulk of what most people think of as fishes belong. Marine fishermen tend to divide marine fish into white fish – those with white flesh, such as flatfish and members of the Cod Family – and those with brown or pink flesh, such as Herring, Mackerel and Salmon. Another division is between demersal species, which live on or near the sea-bed, and pelagic species, which swim near the surface. Freshwater fish are divided by anglers into gamefish (Salmon, Trout, Grayling and other members of the Salmon Family) and coarse fish, comprising all the rest.

REPTILES Reptilia

Common Lizard *Lacerta vivipara*. Common on heaths, moors, dunes and other warm, dry places on light soils almost throughout Britain and Ireland. It is the only true reptile found in Ireland. Food is mainly insects and spiders. The young are born alive, not as eggs, from June to August. The period from mid October to late February is spent in hibernation. 10–16 cm.

Sand Lizard *Lacerta agilis*. A very local species, virtually confined to sandy heaths on the Surrey–Hampshire border, in the New Forest and Dorset, and to the coastal sand-dunes of S Lancashire. Threats to its habitats have made it one of Britain's most endangered species. It hibernates for slightly longer than the Common Lizard. The young hatch from eggs laid in shallow holes. 15–20 cm.

Slow Worm *Anguis fragilis*. This legless lizard is often mistaken for a snake, but it is quite harmless. Found throughout the mainland of Great Britain, it is commoner in the S. It frequents drier habitats, heaths, commons, gardens, hedgebanks and open woods, and hibernates from mid October to the end of February. A form with blue spots is sometimes found. 30–50 cm.

Grass Snake *Natrix natrix*. Widespread in England and Wales, it is completely harmless. It is easily recognised by the pale ring on its neck. It hibernates from late October to the end of February. The young hatch from eggs, often laid in manure heaps and other warm places. 70–150 cm; The largest-ever British specimen was 1.75 m long.

♀

Adder *Vipera berus*. Our only **poisonous** snake and the only snake in Scotland is found throughout mainland Great Britain in dry, open country. Its venom only rarely kills humans. It is easily told by the dark zigzag stripe down its back. Hibernation is from mid October to late February. The young are born, from eggs that hatch immediately, in August and September. 50–60 cm.

♂

poisonous

Smooth Snake *Coronella austriaca*. The rarest of our 3 native snakes has much the same very local distribution as the Sand Lizard, on which it feeds, being confined to sandy heathlands from Surrey to Dorset, including the New Forest. Like our other snakes, it hibernates in the winter. The young begin to emerge from the eggs as soon as they are laid. 55–75 cm.

AMPHIBIANS Amphibia

Smooth Newt *Triturus vulgaris*. The commonest of our 3 newts is widespread on the mainland of Great Britain, but more local in Ireland, where it is the only newt. It lives in ponds in spring and summer, but hibernates on land from mid October to February. In autumn and winter the bright orange–red on the underparts becomes much duller. 7–9 cm.

Palmate Newt *Triturus helvetica*. The rarest of our newts is widespread on the mainland of Great Britain, but completely absent from Ireland. It is commoner in hill country, though also found in the lowlands. It has a much less marked crest along its back than the Smooth Newt, and its distinctive webbed feet give it its English name. 8–9 cm.

Great Crested Newt *Triturus cristatus*. The largest European newt, it is widespread on the mainland of Great Britain, but less frequent than the Smooth Newt. It is absent from Ireland. It has a much larger and more conspicuous crest than the Smooth Newt and also has a distinctively warty skin. It is highly distasteful to potential predators. 12–16 cm.

Common Frog *Rana temporaria*. Widespread and common, although decreasing, on the mainland of Great Britain, it is much scarcer in Ireland, where it is not native, at any rate in the E. Frogs hibernate in the mud by or at the bottom of ponds from mid October to the end of February, and emerge to lay their eggs (frogspawn) in different ponds in March and April. 7–9 cm.

frogspawn

Common Toad *Bufo bufo*. Widespread and common on the mainland of Great Britain, this toad is absent from Ireland. It hibernates on land, in much drier places than the Frog, from mid October to mid March, emerging to spawn in April. The eggs are in strings, not clusters. Toads are extremely distasteful to potential predators, unless these are able to tear the Toad's skin off unscathed. 8–12 cm.

toadspawn

Natterjack Toad *Bufo calamita*. The rarest of our 3 native frogs and toads is now an officially endangered species, with a limited distribution, mainly in E, S and NW England, SW Scotland, and SW Ireland, where it is apparently native. It has a shorter hibernation period than the Common Toad, from November to late February, and spawns from mid April to June. 6–8 cm.

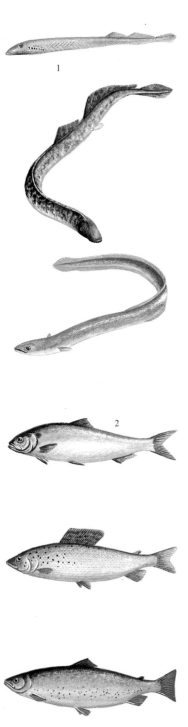

CYCLOSTOMES Cyclostomata. A primitive group that some zoologists no longer regard as fish.

Lampern or **River Lamprey** *Lampetra fluviatilis* (1). An eel-like fish that sucks blood from other fish with its sucker-like jaws, or eats carrion. In autumn it leaves the sea and ascends rivers to spawn on gravelly bottoms in the spring, after which it dies. Up to 40 cm. The smaller **Brook Lamprey** *L. planeri* spends all its life in small rivers or streams. Up to 16 cm.

Sea Lamprey or **Stone-sucker** *Petromyzon marinus*. Much larger than the River Lamprey, it has a similar life history, except that it does not ascend rivers until the spring. This is the lamprey eaten with relish by the rich in the Middle Ages. Up to 75 cm.

EEL FAMILY Anguillidae
Eel *Anguilla anguilla*. A snake-like fish with an extraordinary life history. Born in the Sargasso Sea in the W Atlantic, young eels swim across to ascend British rivers at 3 years old, as elvers. They spend 5–6 years in our rivers as yellow eels before changing to silver eels and returning to the sea to breed or die – it is not yet known which. Up to 100 cm.

HERRING FAMILY Clupeidae. Slim, shoaling fish, frequently migratory, most of them living in the sea.
Allis Shad *Alosa alosa* (2). Shads are herring-like fish that live in the sea and enter rivers to breed in the spring. The Allis Shad (up to 70 cm) usually stays in the estuaries to breed, but its close relative the **Twaite Shad** *A. fallax* (up to 55 cm) goes up-river. At one time there was an important Allis Shad fishery in the Thames.

SALMON FAMILY Salmonidae. Mostly well-known gamefish.
Grayling *Thymallus thymallus*. A wholly freshwater relative of the Trout, identified by the long fin on its back. It prefers fast-flowing streams and is common in England and Wales, but introduced in Scotland and absent from Ireland. Up to 60 cm.

Char *Salvelinus alpinus*. Another relative of the Salmon and Trout, but non-migratory and usually landlocked in deep, cold mountain lakes in Scotland, Ireland and the English Lake District. Some of the more distinct forms have been given special names, such as the Torgoch of Snowdonia and Merioneth, and the Haddy of Loch Killin, Invernessshire. Up to 80 cm.

Salmon *Salar salar.* A large gamefish that spawns in the upper reaches of many larger rivers of Scotland, Ireland, Wales and W England, returning as kelts to the sea, where many die, but some return to spawn again. The young fish are called first alevins and then parr; after 2 years they turn into silvery smolts and migrate to the sea, whence they return as grilse to spawn. Up to 150 cm.

Trout *Salar trutta.* A gamefish, smaller than the Salmon, with 2 forms, Sea or Salmon Trout that spawn in rivers but return to the sea, and Brown or River Trout that spend all their lives in freshwater rivers or lakes. Young Trout are also called parr, and young Sea Trout become silvery smolts like Salmon. Mature Trout average 10–15 kg in weight. 80–100 cm.

Rainbow Trout *Salar gairdnerii.* A N American fish that is widely introduced into British rivers and lakes by anglers, but rarely becomes established and breeds. It can stand more polluted waters than the Brown Trout and also higher temperatures. It owes its name to the iridescent purple band along its side. Up to 70 cm.

Powan *Coregonus clupeoides.* One of a group of small, herring-like relatives of the Salmon and Trout, colectively known as Whitefish. This species is found in Lochs Esk and Lomond in Scotland, and also in Lake Bala in N Wales, where it is called gwyniad, and in Haweswater and Ullswater in the Lake District, where it is called schelly. 10–70 cm.

Pike *Esox lucius.* An extremely aggressive, predatory fresh-water fish, widespread in British lakes and slow-flowing rivers, that preys on Perch, Roach, Rudd, Dace and other smaller fresh-water fish, as well as various crustaceans and other invertebrates, and even young Moorhens, ducklings and Water Voles. Young Pike are known as jacks. 25–100 cm.

CARP FAMILY Cyprinidae. Favourites of the fresh-water angler.
Carp *Cyprinus carpio.* A large, long-lived – to 40–50 years or more – fresh-water fish, much valued for rearing in ponds, and believed to have been originally introduced to Europe from Asia by the Romans and to Britain in the Middle Ages. Golden Carp, Prussian Carp and Mirror Carp are all varieties bred in captivity, but Crucian Carp (p.196) is a distinct species. 25–100 cm.

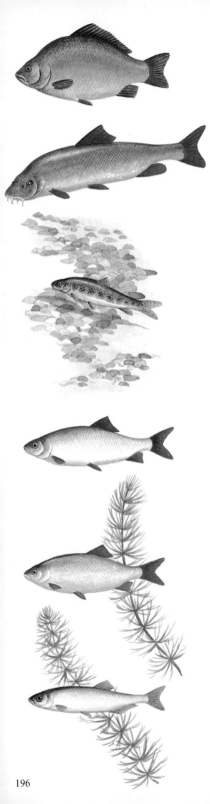

Crucian Carp *Carassius carassius*. A relative of the Goldfish *C. auratus*, native to Europe but probably introduced in England, where it is commonest in the Thames Basin and E Anglia. The curious name is derived from its German name, *Karausche*, latinised as *Carassius*. Hybrids with the common Carp are sometimes raised in fish farms and stocked in small waters. 10–45 cm.

Barbel *Barbus barbus*. Our rarest native fish, which has always been confined to rivers flowing into the North Sea, and is now mainly found in the Thames and Trent. It is named from the 4 barbels, or fleshy filaments, hanging from its mouth, though these are also found in the common Carp. It spends the winter in a torpid condition in deeper water. 30–90 cm.

Gudgeon *Gobio gobio*. A small, bottom-living fresh-water fish, not usually more than about 14 cm in length, frequent in rivers, streams and ponds in England, Wales and Ireland, but absent from Scotland. In summer small shoals swim in shallow water, where they spawn; in winter they are still active, but in deeper water. 8–20 cm.

Roach *Rutilus rutilus*. A widespread and common favourite of the angler, found in both still and running fresh water throughout England and Wales, except for the western extremities, and in Scotland. In Ireland it is rare, and the Rudd, which is common, is confusingly called Roach. It is usually, but not always, more silvery and less ruddy than the Rudd. 10–25 cm.

Rudd *Scardinius erythrophthalmus*. More local in England and Wales than the Roach, it is absent from Scotland, but commoner in Ireland. It has a deeper body than the Roach and is usually more rufous, especially on the fins. A promiscuous fish, it is noted for shoaling and hybridising with other closely related fish, such as Roach, Bleak, Bream and White Bream. 20–35 cm.

Bleak *Alburnus alburnus*. A small relative of the Carp, it lives in shoals along the edges of lakes and slow-moving rivers, where it feeds on gnats and their larvae and water fleas and often leaps out of the water. Widespread in England and Wales, except for the Lake District, W Wales and the southern coastal counties, it is absent from Scotland and Ireland. 12–20 cm.

Minnow *Phoxinus phoxinus*. The smallest member of the Carp Family, widespread and common in lakes, ponds, brooks and streams throughout mainland Great Britain, except for the far N of Scotland; in Ireland it is very local. It is a favourite bait of anglers. In winter Minnows retire into deep water and hide under stones or in holes in the banks. Up to 12 cm.

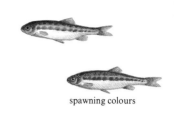

spawning colours

Dace *Leuciscus leuciscus*. A widespread and common surface-living river fish, preferring fast streams and much favoured by coarse fishermen. It is found in most English and Welsh rivers, but is absent from Scotland and Ireland. A small fish, it rarely exceeds 25–30 cm.

Chub *Squalius cephalus*. Distinctly larger than the Dace (above), sometimes attaining twice its size, but with similar habits, and equally sought after by coarse fishermen. On hot summer days shoals may be seen near the surface, sometimes jumping up after flies. Found in most rivers S of the Firth of Forth, but absent from the Highlands and from Ireland. Up to 60 cm.

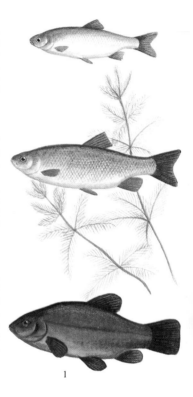

Tench *Tinca tinca* (1). A widespread and frequent sluggish fresh-water fish of muddy lakes and ponds on the British mainland as far N as Loch Lomond. Like Carp, it can live for some time out of the water, or in mud at the bottom of dried-up ponds. The **Golden Tench** is an ornamental form, like a large Goldfish. Up to 50 cm.

1

Common Bream, Carp Bream or **Bronze Bream** *Abramis brama*. A fresh-water fish of lakes and slow-flowing streams that is especially common in the Norfolk Broads and parts of Ireland. On the British mainland it is not found N of Loch Lomond. Spawning takes place on aquatic vegetation in May. Up to 60 cm.

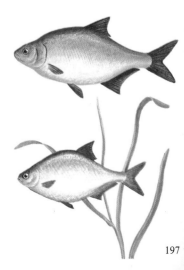

Silver Bream or **White Bream** *Blicca bjoernka*. A much more local fish than the Common Bream, being more or less confined to the rivers flowing into the North Sea from Yorkshire to Suffolk. It is also found in lakes and sometimes swims in shoals with Roach and Rudd. A bottom feeder, it attaches its eggs to waterweeds, like the Common Bream. Up to 35 cm.

197

LOACH FAMILY Cobitidae. A family of bottom-living fish with barbels around the mouth.

Stone Loach *Nemacheilus barbatula*. A widespread and frequent small fish, not more than 10–12 cm in length, found in clear streams with gravelly bottoms throughout Great Britain and Ireland, except in the N Scottish Highlands. It spawns in April and May, and feeds on small worms, shrimps and insect larvae.

Spined Loach *Cobitis taenia*. A smaller (5–10 cm) and much more local relative of the Stone Loach, not having been recorded from Wales, Scotland or Ireland. It has a pair of spines on its head which serve as weapons, and if handled it will use them. It often lies still on the bottom, with only its head protruding from the sand or stones.

PERCH FAMILY Percidae
Perch *Perca fluviatilis*. A strikingly barred, gregarious fresh-water fish, widespread and common in rivers, lakes and ponds throughout Great Britain and Ireland, except for parts of the Scottish Highlands. It grows to about 45 cm, and spawns in shallower water between March and May.

Ruffe or **Pope** *Acerina cernua*. A small (to 20 cm), gregarious fresh-water fish, allied to the Perch, that is rather local in still or slow-moving waters in England and Wales as far N as Lancashire and Yorkshire. The name Ruffe refers to the roughness of its scales; how it came also to be called Pope is not known.

BULLHEAD FAMILY Cottidae
Bullhead or **Miller's Thumb** *Cottus gobio*. A small (10–18 cm), spiny, broad-headed fresh-water fish, with 3 marine relatives. It is widespread and frequent in England and Wales, but absent from Scotland and Ireland. A solitary fish, it lies under stones in clear brooks or shallow lakes, and if handled can wound with its spines.

STICKLEBACK FAMILY Gasterosteidae
Three-spined Stickleback or **Tittlechat** *Gasterosteus aculeatus* (1). A widespread, common, gregarious, small minnow-sized fish (5–8 cm) of fresh or brackish water throughout Great Britain and Ireland, distinguished by its 3 spines from the local and southern **Ten-spined Stickleback** *Pygosteus pungitus* and the larger and exclusively coastal **Fifteen-spined Stickleback** *Spinachia spinachia*.

1

♂ in spawning colours

198

Butterflies and Moths

Lepidoptera, the order of insects (see p.219) containing the butterflies and moths, are characterised by the powdery scales on their wings, which often make brightly coloured patterns. This is the main factor which makes the butterflies one of the few insect groups that are attractive to the general public. They pass through 3 other life stages before emerging as perfect insects, beginning as eggs, which hatch into larvae known as caterpillars, and then becoming pupae or chrysalids, a term deriving originally from the colourful pupae of certain butterflies.

insects as the tortoiseshells, the Peacock, the admirals and the fritillaries; the Lycaenidae, including the blues, coppers and hairstreaks; the Pieridae, comprising the whites and the Brimstone; and the Hesperiidae, which consists of the skippers.

Moths, of which there are some 2,120 British species, form 3 sub-orders, of which the Heteroneura contains the great majority. The old division between Macrolepidoptera and Microlepidoptera has now been discarded – indeed there was always an overlap in size between them. What were formerly called

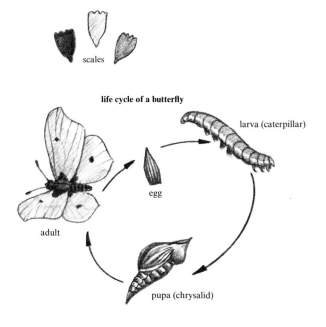

scales

life cycle of a butterfly

larva (caterpillar)

egg

adult

pupa (chrysalid)

Butterflies belong to the superfamily Papilionidae, and are characterised by their club-tipped antennae and by not having a mechanism for hooking the fore and hind wings together, which moths possess. All butterflies, but only a few moths, fly by day. There are some 68 British species, of which 10 are only rare or occasional visitors, in 8 families. The most important of these are the Saturnidae, the browns; the Nymphalidae, including such brilliantly coloured

'micros' are now mainly put in 5 super-families of the Heteroneura. Some of the most important moth families are the Sphingidae or hawk-moths, which have large caterpillars often carrying a 'horn' at one end; the Noctuidae, mostly rather drab in appearance but containing such attractions as the yellow underwings; and the geometers, a group with 'looper' caterpillars (see p.218), formerly all included in the Geometridae, but now split into 6 families.

FRITILLARY FAMILY Nymphalidae

Small Tortoiseshell *Aglais urticae*. One of the commonest British butterflies, appearing almost throughout the summer, from March to May and again from July to October, when it hibernates. The foodplant of the blackish brown caterpillar is the stinging nettle *Urtica dioica*. Widespread throughout the British Isles, but less common in Scotland and Ireland. WS *c*. 5 cm.

Painted Lady *Vanessa cardui*. A butterfly that immigrates from the Mediterranean in greatly varying numbers from year to year. Most often seen from July to September, it sometimes also appears in spring. Caterpillars resulting from eggs laid by spring visitors are sometimes found on various kinds of thistle, *Cirsium* and *Carduus*. A return migration can be seen in autumn. WS *c*. 5.5 cm.

Red Admiral *Vanessa atalanta*. Like the Painted Lady, most British specimens of the much commoner Red Admiral arrive from the Mediterranean in spring or summer. Some may return and a very few are able to hibernate through mild winters. The caterpillars feed largely on Stinging Nettle *Urtica dioica*. The adults frequent many garden plants, but especially the Butterfly Bush *Buddleia davidii*. WS 6.2 cm.

Peacock *Inachis io*. A common butterfly, with much the same double-brooded and hibernating pattern as the Small Tortoiseshell. Both species habitually take refuge in buildings in late summer and autumn and come out during mild winter spells. The blackish caterpillars feed on Stinging Nettle *Urtica dioica*. Even less common in Scotland and Ireland than the Small Tortoiseshell. WS 5.5 cm.

White Admiral *Ladoga camilla*. A local butterfly, whose southern English range expanded greatly over the Midlands during the 1930s and 1940s, but has now contracted again. Its green caterpillars feed only on Honeysuckle *Lonicera periclymenum*, and the handsome adults fly in woodland in July and August, when they like feeding on the nectar from Bramble blossom *Rubus fruticosus*. WS 5.5 cm.

Purple Emperor *Apatura iris*. One of the most handsome but least often seen British butterflies, thanks to its habit of frequenting the tops of oaks and other trees. It is a very local insect, being almost confined to extensive woodlands in Hampshire, Wiltshire and some other central S English counties. The caterpillars feed on Sallow or Goat Willow *Salix caprea*. WS 6.2–6.8 cm.

Comma *Polygonia c-album*. A striking butterfly looking somewhat like a small Tortoiseshell with ragged wings. The white comma-shaped mark which gives it its name is on the hind underwing. Not uncommon in the S, but thinning out northwards and not N of the Humber–Mersey line. The caterpillars are fairly omnivorous, but most often eat Stinging Nettle *Urtica dioica*. Flying spring and autumn. WS 4.4 cm.

Pearl-bordered Fritillary *Boloria euphrosyne* (1). Fritillaries are distinguished by their black and warm brown upperparts and silver spots beneath; their caterpillars usually feed on various kinds of violet. The Pearl-bordered is widespread, but mainly in the S; flying in May and early June. WS 3.7 cm. The slightly smaller **Small Pearl-bordered Fritillary** *B. selene* does not appear till June.

Dark Green Fritillary *Argynnis aglaia*. One of the more easily identifiable larger fritillaries, because of the greenish tinge on the underwing. Frequents open country, such as downs, heaths and dunes throughout Britain and Ireland, but mainly in the S; flying in July and August. Like other fritillaries, its caterpillar hibernates through the winter. WS 5.5–6.2 cm.

High Brown Fritillary *Argynnis adippe*. Much more local than the Dark Green Fritillary, and mainly in woodland rides and along the edges of woodland; mostly in S England, and not extending into either Scotland or Ireland. Has chestnut spots instead of a green wash on its underwing. Flies in July and August. WS 5.5–6.2 cm.

Silver-washed Fritillary *Argynnis paphia*. Much commoner than the High Brown Fritillary in its woodland haunts, especially in S and SW England, the New Forest being one of its favourite localities; not N of the Midlands, but is scattered over Ireland. Easily told from the High Brown by the more blurred appearance of the silver markings in the underwing. Flies from July to September. WS 6.2 cm.

Marsh Fritillary *Euphydryas aurinia*. One of the less common small fritillaries, with underwing markings buffish rather than silver. Local in S England, the W side of Great Britain, and the N half of Ireland, inhabiting various kinds of damp ground where its food plant, Devil's Bit Scabious *Succisa pratensis*, grows. Flies in late May and early June. WS 3.7 cm.

BROWN FAMILY Satyridae

Speckled Wood *Pararga aegeria.* A common woodland butterfly of S England, more local in Wales, the Midlands and Ireland, and scarcely extending N of the Mersey–Humber line. The caterpillars feed on various grasses, and there are 2 broods, the adults flying from April to early June and again from July to early September. WS 4.4 cm.

Wall or **Wall Brown** *Lasiommata megera.* A common grassland butterfly, favouring hot, sunny sites where it can bask on walls, rocks or bare ground. Common in England and Wales, but thinning out northwards and not extending into Scotland; rather scarce in Ireland. The caterpillars feed on grasses, and the adults fly in May and June and again from late July to mid September. WS 4.4 cm.

Scotch Argus *Erebia aethiops* (1). The commoner of our 2 mountain butterflies, favouring woodland borders from the Lake District northwards. The caterpillars feed on Purple Moor-grass *Molina caerulea*, and the adults fly from mid July to early September. WS 3.7 cm. **Mountain Ringlet** *E. epiphron* is confined to the Lake District and the central Highlands, flying in June and July. WS 3.1 cm.

Marbled White *Melanargia galathea.* Our only butterfly which is strikingly chequered black and white, it is common in chalk and limestone grassland in S and Midland England, with an outpost on the chalk of the Yorkshire Wolds. Its caterpillars feed on Sheep's Fescue *Festuca ovina* and other grasses, and the adults fly in July and August. WS 4.4–5 cm.

Grayling *Hipparchia semele.* A common butterfly of open habitats, heaths, downs, coastal cliffs and dunes, most often seen not far from the sea in England and Wales; rather scarce in Ireland. The caterpillars feed on a variety of grasses and the adults are on the wing from late July to early September. WS 5–5.6 cm.

Ringlet *Aphantopus huperantus.* The darkest of our brown butterflies is widespread and common in shady places, alongside woods and hedgerows as far N as mid Scotland, but is commonest in the S; rather local in Ireland. The caterpillars feed on various grasses and the adults fly from late June to August. WS 4.4 cm.

Gatekeeper *Pyronia tithonus*. A common butterfly of pastoral farmland, hedgerows and scrub in S and Midland England and coastal Wales, thinning out northwards and rare N of the Humber–Mersey line and in Ireland. The caterpillars feed on various grasses, and the adults fly from mid July to mid August. WS 3.7–4 cm.

Meadow Brown *Maniola jurtina*. Generally the commonest of our brown butterflies, and the most widespread, though fairly local in Scotland and Ireland. The caterpillars feed on grasses of various kinds and the adults fly from mid June to mid August in all kinds of grassy places, meadows, pastures, downs, heaths, sea cliffs, road verges, parks and large gardens. WS 4.4–5 cm.

Large Heath *Coenonympha tullia*. A local butterfly, dependent on the distribution of its caterpillar's food-plant, White Beaksedge *Rhynchospora alba*, which grows on acid peat. The Large Heath therefore occurs only on peat bogs and mosses, and nearby moors and mountains; it is not now found S of mid Wales and Cheshire and is very rare in Ireland. Adults fly from mid June to mid July. WS 3.2 cm.

Small Heath *Coenonympha pamphilus*. A very common butterfly in the Midlands and S England, thinning out considerably northwards and in Ireland. Its caterpillars feed on a variety of fine-leaved grasses, and the adults fly from May to September, in 2 broods, in all kinds of rough, grassy places, such as heaths, downs and dunes. WS 2.5 cm.

DUKE OF BURGUNDY FRITILLARY FAMILY Nemeobiidae
Duke of Burgundy Fritillary *Hamearis lucina*. Superficially like the other fritillaries, the Duke of Burgundy is quite unrelated to them, being much smaller and with no silver spots beneath. A local insect of the S half of England, it is most often found where its foodplants, Cowslip and Primrose, grow, along wood edges and in half-shade. The adults fly in May and June. WS 2.5 cm.

WHITE FAMILY Pieridae
Brimstone *Gonepteryx rhaemni*. The male is our only all-yellow butterfly, but females are so pale yellow that they are often mistaken for Large Whites. Very common SE of the Humber–Severn line, it is scarce in Wales, N England and Ireland, and absent from Scotland. The caterpillars feed on Buckthorn and Alder Buckthorn, and the adults fly from March to October, hibernating in winter. WS 5.6 cm.

Large White *Pieris brassicae*. One of our 2 cabbage whites, whose black-speckled caterpillars are unwelcome visitors on cabbages and cauliflowers. All too common in the S, it is less of a problem to Scottish and Irish gardeners. Our native stock is augmented by many immigrants from the Continent, and adults may be seen from May to September or even October. WS 6 cm.

Small White *Pieris rapae*. The smaller and commoner of the 2 cabbage whites, whose green caterpillars are even more frequent in the kitchen garden. Has roughly the same distribution as the Large White and can be seen almost anywhere, but usually in or near towns, villages or farms. Adults have 2 broods, April–May and June–August. WS 4.3 cm.

Green-veined White *Pieris napi*. Commoner than the 2 cabbage whites in the countryside generally, especially in woods and the wilder habitats; its caterpillars feed mainly on Garlic Mustard *Alliaria petiolata*, Cuckooflower *Cardamine pratensis* and other wild members of the Cabbage Family. Widespread and common, but less so in the N and Ireland. Adults fly in 2 broods, April–June and July–August. WS 4–4.5 cm.

Orange-tip *Anthocharis cardamines*. The male is one of our handsomest and most distinctive butterflies, but the grey and white female looks like a slender cabbage white. Widespread and common in woods and hedgerows S of the Humber–Mersey line, it thins out northwards and in Ireland. The caterpillars feed on the same 2 plants as the Green-veined White; the adults fly from April to July.

Clouded Yellow *Colias crocea* (1). A migrant from the Continent, much commoner in some years than others, and most often seen in the S from July to October. Early arrivals may breed, the caterpillars feeding on clovers, Lucerne and other members of the Peaflower Family, but the resulting adults do not overwinter. WS 5 cm. The **Pale Clouded Yellow** *C. hyale* also occurs, but is much scarcer. WS 4.5 cm.

Wood White *Leptidea sinapis*. A very local insect of southern and a few Irish woodlands, it differs from our other white butterflies especially in its small size and wavering, uncertain flight. The caterpillars feed on Birdsfoot Trefoil *Lotus corniculatus* and other members of the Peaflower Family, and the adults are on the wing from late May to July. WS 3.5 cm.

BLUE FAMILY Lycaenidae

Small Blue *Cupido minimus*. One of the smaller and less blue of our native blue butterflies, only the male having a bluish dusting on the upperside; neither sex has any orange spots. Rather local on chalk and limestone grassland, mainly in the S, the caterpillars feeding on Kidney Vetch *Anthyllis vulneraria*. The adults fly in May and June, and sometimes again in August. WS 2.2 cm.

Silver-studded Blue *Plebejus argus*. Another local butterfly, largely confined to heathlands S of the Thames, but occasionally further N and on downs and dunes. The male's dark border separates it from the Common Blue. The caterpillars feed on Gorse, Broom and some other members of the Peaflower Family, and the adults fly from mid June to mid August. WS 2.5 cm.

Common Blue *Polyommatus icarus* (1). Generally the commonest and most widespread of our blue butterflies, but much less so in the N and in Ireland. It frequents all kinds of open grassy and heathy places, its caterpillars feeding on clovers and other peaflowers. The adults fly from May to September, with 2 broods. WS 2.8 cm. The very similar **Adonis Blue** *Lysandra bellargus* is now scarce on chalk in the S.

Chalkhill Blue *Lysandra coridon*. A most distinctive pale blue butterfly, whose colonies are locally frequent on chalk and limestone grassland in the S. Its caterpillars feed on Horseshoe Vetch *Hippocrepis comosa* and several other members of the Peaflower Family. The adults fly in July and August. WS 3.1 cm.

Holly Blue *Celastrina argiolus*. The only blue butterfly at all frequent in woods and gardens, its caterpillars feed on the flower-buds of Holly and Ivy. Locally frequent in the Midlands and S, but rare elsewhere; the adults fly from April to June and again in August and September. The female is also blue, unlike most blue butterflies; there are no orange spots. WS 2.8 cm.

Brown Argus *Aricia agestis*. Our only 'blue' in which both sexes are brown. Locally frequent in grassy places in S and Midland England, the caterpillars feeding on Storksbill *Erodium cicutarium* and Rock-rose *Helianthemum chamaecistus*. The adults fly in May and June and from July to September. The Scottish and N of England species *A. artaxerxes* has a white spot on its forewing. WS 2.5 cm.

Small Copper *Lycaena phlaeas*. A most attractive small butterfly, quite common in all kinds of open, grassy places, but much less so in the N and in Ireland. The caterpillars feed mainly on Sheep's Sorrel *Rumex acetosella*. There are 3 broods, the adults flying from May to September. Exceptionally there is a fourth emergence, in October. WS 2.5 cm.

Green Hairstreak *Callophrys rubi*. Our only butterfly with a bright green underside is widespread throughout Britain and Ireland, but much more frequent in S and Midland England. It flies in May and June in all kinds of places with half-shade, woodland edges, scrub, downs and heaths. The caterpillars feed on an unusually wide variety of plants, often on Gorse. WS 2.5 cm.

Brown Hairstreak *Thecla betulae*. A very local butterfly of woods, scrub and other places where the food-plant of its caterpillars, Blackthorn, grows. Most localities are in S England, but a few also in the Midlands and Wales. The adults fly in August and September. WS 3.1 cm.

Purple Hairstreak *Quercusia quercus*. A locally frequent butterfly of high Oak forest, in whose tops it spends most of its time, so that it is not often seen. Mainly in the S half of England and Wales; very rare in Scotland and Ireland. The caterpillars feed on Oak leaves, and the adults fly from July to September. WS 3.1 cm.

White-letter Hairstreak *Strymonidia w-album*. Another local butterfly almost confined to Midland and S England, where it flies in July and August wherever there are elms, especially Wych Elm. The recent Dutch Elm Disease epidemic must inevitably threaten its status here. It has no orange on the forewings, like the Brown and Black Hairstreaks. WS 2.8 cm.

Black Hairstreak *Strymonidia pruni*. Our rarest hairstreak, confined to a series of woods that contain dense thickets of old Blackthorn, its caterpillar's food-plant, and stretch in a narrow band from near Oxford to near Peterborough, Northants. It has become much scarcer with the decrease in its habitat in recent years, and is now one of our rarest butterflies. WS 2.8 cm.

Dingy Skipper *Erynnis tages*. An unassuming little butterfly, widespread and quite frequent in England and Wales, though thinning out northwards, and scarce in Scotland and Ireland. It frequents many kinds of grassy and heathy places, because its caterpillar's food-plant, Birdsfoot Trefoil *Lotus corniculatus*, is very catholic in its choice of soil. Adults fly in May and June. WS 2.5 cm.

Grizzled Skipper *Pyrgus malvae*. Smaller than the Dingy Skipper and easily told by its whitish blotches, it is more restricted in range, being confined to England S of the Humber–Mersey line, and S Wales. The habitat is very similar, although the caterpillars feed on Wild Strawberry *Fragaria vesca* and other members of the Rose Family. Adults fly in May and June. WS 2.2 cm.

Small Skipper *Thymelicus sylvestris* (1). A common butterfly of grassy places – its caterpillars eat various grasses in England and Wales – though much commoner in the S and the Midlands. Adults fly in July and early August. WS 2.2–2.5 cm. The very similar **Essex Skipper** *T. lineola*, distinguished by the black tips to the underside of its antennae, is found only in SE England. WS 2.5 cm.

Silver-spotted Skipper *Hesperia comma* (2). A very local butterfly of chalk grassland in SE England; its caterpillars feed on Sheep's Fescue *Festuca ovina* and other grasses. The adults fly in August and early September. WS 3.4 cm. Another very local butterfly is the **Lulworth Skipper** *Thymelicus acteon*, confined to grassy cliffs in Dorset and E Devon. It resembles the Small Skipper, but is a muddier brown. WS 2.5 cm.

Large Skipper *Ochlodes venata*. Generally the commonest as well as the largest skipper, extending N to Cumbria; it is the earliest to appear, flying in June and July in all kinds of grassy places, usually with some bushes, where its caterpillars are able to feed on various grasses. Like all skippers, it sometimes holds its wings along its body like a moth. WS 3.4 cm.

Chequered Skipper *Carterocephalus palaemon*. Now our rarest skipper, and indeed one of our rarest butterflies, it is found only in W Inverness-shire, having become extinct in its former strongholds in and around N Northants. The caterpillars feed on various grasses, such as False Brome *Brachypodium sylvaticum*, and the adults fly in woodland clearings in May. WS 2.5 cm.

207

RARITIES AND VISITORS

Large Blue *Maculinea arion* (1). Became extinct in 1979, but formerly inhabited grassy places with anthills in Devon, Cornwall and the Cotswolds. WS 3.1 cm.

Camberwell Beauty *Nymphalis antiope* (2). An occasional migrant from N Europe, first recorded at Camberwell in S London. WS 6.2–7.5 cm.

Swallowtail *Papilio machaon* (3). Now confined to the Norfolk Broads. Its caterpillars feed on Milk Parsley, and the adults fly May–June and again August–September. WS 6.9–7.5 cm.

Monarch or **Milkweed Butterfly** *Danaus plexippus* (4). Occasionally strays here from N America or from the Canary Islands, where it also breeds. WS 8 cm.

Large Copper *Lycaena dispar* (5). Became extinct a century ago, when the fens were drained, but a colony has been reintroduced at Woodwalton Fen, Huntingdonshire, where its caterpillars feed on Great Water Dock. WS 3.1–3.7 cm.

Bath White *Pontia daplidice* (6). A rare immigrant. WS 4.4 cm.

Ghost Moth or **Ghost Swift** *Hepialus humuli*.
Widespread and frequent throughout Great
Britain and Ireland; the pure white males
are a striking sight as they weave back and
forth over grassy places on summer nights,
often appearing to be manipulated by some
invisible puppet-master. The whitish caterpillars feed on the roots of grasses and
other plants and are regarded as a crop
pest. WS 5.5–6 cm.

Common Swift *Hepialus lupulina*. A widespread and common moth throughout Great
Britain, but rather local in Ireland, it often
comes to light. It may be seen flying low
over grassland in the June dusk. The caterpillars feed on the roots of various grasses.
WS 3.5 cm.

Goat Moth *Cossus cossus* (1). Widespread
but local throughout Great Britain and Ireland, this moth owes its name to the smell
emitted by its large (up to 10 cm) caterpillar,
which spends its 3- or 4-year life burrowing
in the wood of willow, elm, ash and other
trees. The moth flies in June and July. WS
6.5 cm. **Leopard Moth** *Zeuzera pyrina* (2) is
a widespread black and white relative of
the Goat Moth. It flies in July and August
and is nocturnal. The larvae tunnel into
hardwood trees and are a great pest in
orchards. WS 6.5 cm.

Currant Clearwing *Aegeria tipuliformis*. One
of the commoner of a group of moths that
look confusingly like small Hymenoptera
(wasps). Widespread in England, it is much
more local in Wales, Scotland (only in the
S) and Ireland, being most often seen resting on the foliage of currant and other
garden bushes in June or July. The caterpillars feed inside the stems and shoots of
currant and gooseberry bushes. WS 1.8 cm.

Six-spot Burnet *Zygaena filipendulae*. A
common day-flying moth of grassy places,
where its distinctive straw-coloured cocoons
can often be seen attached to grass stems.
It is found throughout Great Britain and
Ireland, especially on the downs and dunes
where its caterpillar's (p.218) favourite food-
plants, clovers, trefoils and other small peaflowers, grow. WS 3.2 cm.

The Forester *Procris statices*. A widespread
moth of damp meadows in England, it is
local in Scotland, Wales and Ireland. The
adult is especially fond of visiting Ragged
Robin flowers in June, but the caterpillar
feeds on Common Sorrel. WS 3 cm.

larva

Green Oak Moth *Tortrix viridana* (1). This is the tiny moth whose caterpillars devastate Oak trees in early summer. WS 1 cm. **Large White Plume Moth** *Alucita pentadactyla* (2) is the largest and commonest of the plume moths, widespread in England, Wales and Ireland. Its caterpillars feed on Bindweed. WS 1.5 cm.

Small Magpie Moth *Eurrhypara hortulata.* This pyralid moth is common in Ireland and England, more local in N England and S Scotland. It flies in June and July, and the caterpillars can be found rolled up inside nettle leaves a few weeks later. WS 2.6 cm.

Mother of Pearl Moth *Sylepta ruralis.* Another pyralid moth, common where nettles grow throughout Great Britain and Ireland. The caterpillars may be found rolled up in nettle leaves in the spring, the moth emerging in July. WS 2.2 cm.

Grass Moth *Crambus pratellus.* One of the commonest of a large group of moths which is found in grassy places throughout Great Britain and Ireland and even in the N and W isles. The moth flies throughout the summer, and its caterpillar feeds on grasses, especially Tufted Hair-grass *Deschampsia caespitosa*. WS 1.5 cm.

Small Ermine Moth *Yponomeuta padella.* This is the small moth whose tiny, black-spotted caterpillars form large webs on Hawthorn and various other shrubs (p.218). Though superficially similar, it is no relation to the much larger Buff Ermine and White Ermine moths, whose caterpillars are known as 'woolly bears'. WS 2 cm.

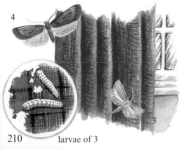

Clothes Moth *Tineola bisselliella* (3). One of the commonest of a small group of moths whose caterpillars attack clothing and other fabrics. WS 1 cm. Another is the **Brown House Moth** or **False Clothes Moth** *Borkenhausia pseudopretella* (4), commoner in many houses, which is slightly larger and dull brown all over. WS 1.5 cm.

Buff-tip *Phalera bucephala*. A widespread and common member of the Prominent Family that occurs throughout Great Britain and Ireland and is especially common in the London suburbs, flying in June or July. Its colonial caterpillars are voracious and will often strip whole tree-branches of their leaves. Limes and elms are favourites, but almost any tree will do. WS 5.5 cm.

Puss Moth *Cerura vinula* (1). A rather dull-looking large moth that is widespread and common throughout Great Britain and Ireland, wherever Poplars or Willows grow. Its remarkable caterpillar (p.218) puts on a formidable threat display with its face mask and tail whips. WS 3 cm. **Poplar Kitten** *C. hermelina* (2) is similar but smaller, and absent from Scotland. WS 5.5 cm.

Swallow Prominent *Pheosia tremula*. One of the commoner members of the large Prominent Family, it is widespread and locally frequent throughout Great Britain and Ireland, except for the far N of Scotland. It appears in May and again in August, and the caterpillars prefer Poplar, but will also eat Sallow. WS 5 cm.

Buff Ermine *Spilosoma lutea* (3). A common and widespread species almost throughout Great Britain and Ireland, it is most familiar in the form of its 'woolly bear' caterpillar, when seen scurrying across a path on its way to pupate. WS 4 cm. **White Ermine** *S. lubricipeda* has a similar caterpillar, but the adult is white, not buff. Both moths emerge in June. WS 4 cm.

Garden Tiger *Arctia caja* (4). Another widespread and common moth, found almost everywhere in Great Britain and Ireland, whose hairy caterpillar (p.218) is called a 'woolly bear'. Most 'woolly bears' will feed on almost any low-growing plant. WS 6 cm. **Wood Tiger** *Parasemia plantaginis* (5) is equally widespread, but found mainly on heaths, moors and downs. WS 2 cm.

Cinnabar Moth *Callimorpha jacobaeae* (6). A very common day-flying moth, found throughout Great Britain and Ireland, except for N Scotland, wherever Ragwort or Groundsel, its caterpillar's (p.218) food-plants, grow. Ragwort plants are often completely stripped of their leaves by these conspicuous caterpillars, whose bright colours show that they are distasteful to predators. WS 4 cm.

Heart and Dart *Agrotis exclamationis*. A widespread and common member of this great family, found throughout Great Britain and Ireland, but only in the SE of Scotland. It flies in June and July, and its caterpillar feeds on Chickweed and various other low-growing plants. WS 3.5 cm.

Grey Dagger *Apatele psi*. Another widespread and common noctuid in England and Wales, but apparently absent from Scotland and Ireland. Its handsome caterpillar (p.218) feeds on a variety of trees and shrubs, shrubs, such as Lime, Hawthorn, Blackthorn and Apple. The moth flies in June. WS 3.5 cm.

Antler Moth *Cerapteryx graminis*. A widespread and common noctuid moth found in all parts of Great Britain and Ireland, but less frequent in the S of England. Its caterpillars can occur in plague proportions on moorland and denude extensive areas of vegetation, while attracting large numbers of Rooks and other predators. WS 3 cm.

Herald *Scoliopteryx libatrix*. A handsome moth that often enters houses and barns to hibernate in late summer and autumn, and reappears in spring. It is widespread and common throughout Great Britain and Ireland. The caterpillar feeds on various kinds of willow and sallow, including Osier. WS 4.5 cm.

Large Yellow Underwing *Noctua pronuba*. The largest and one of the commonest of an attractive group of noctuids that have a broad yellow band on the underwing. It is found throughout Great Britain, N to the Moray Firth, and in Ireland. A familiar moth in houses, where it is attracted to light in the summer, its caterpillars feed on various low-growing plants. WS 5.5 cm.

Red Underwing *Catocala nupta*. One of our largest and most attractive moths, it flies in August and September in the S and E of England, and more sparsely N to Derbyshire and N Wales. The caterpillars feed on poplar and willow, which determine the kind of places where the moth is most likely to be found. WS 7 cm.

Angle Shades *Phlogophora meticulosa* (1). A strikingly patterned noctuid moth, easily mistaken for a dead leaf when at rest, that is widespread and common throughout Great Britain and Ireland. Its caterpillars feed on various low-growing weeds. WS 4.5 cm. The handsome **Green Silver Lines** *Bena fagena* (2) is also widespread S of the Moray Firth and in Ireland, its caterpillars feeding on Oak, Beech and other trees. WS 3.5 cm.

Silver Y *Plusia gamma*. A common migratory moth, often flying by day, and named from the Y-shaped mark on its forewings. It is found in almost all parts of Great Britain, Ireland and their adjacent islands. There are 2 waves of migration, in spring and late summer. The caterpillars feed on a wide variety of low-growing plants. WS 4 cm.

Black Arches *Lymantria monacha*. Widespread but local in England and Wales S of Yorkshire, and in Ireland, this attractive moth flies in late July and August. In the New Forest, Hampshire, one of its favourite haunts, it may be seen resting on tree-trunks. The food-plants of the caterpillar are Oak and various other trees. WS 3.5 cm.

Gold-tail or **Yellow-tail** *Euproctis similis*. An all-white moth found throughout England as far N as Yorkshire and Lancashire. Its handsome caterpillar, known as the palmer worm (p.218), is one of those that should be avoided since its hairs produce a severe rash on some people; it feeds on Hawthorn and the foliage of numerous other trees and shrubs. WS 3 cm.

Vapourer *Orgyia antiqua*. A small moth whose female is wingless and whose caterpillar is adorned with variegated tufts of hairs (p.218). It is widespread and common in Great Britain, especially in and around London, but less so in Ireland. The moths emerge in June. The caterpillars feed on a wide variety of trees and shrubs, including the London Plane. WS 2.5 cm; female L 10 mm.

Oak Eggar *Lasiocampa quercus*. Widespread and common throughout Great Britain and Ireland, this day-flying moth is found on moors and in other rough country. The males fly over the heather in search of the females in early summer. The handsome, hairy caterpillars feed on Heather and a good many other plants. WS 5 cm.

213

Drinker *Philudoria potatoria*. A widespread and frequent moth in all kinds of open country more or less throughout Great Britain and Ireland, more familiar as a large, hairy caterpillar sunning itself in April or May, usually on the grasses on which it feeds. The moth flies in July. WS 5–6 cm.

December Moth *Poecilocampa populi* (1). The only autumn- and winter-flying member of the eggar group, often seen around street lamps between October and Christmas. Widespread but local almost everywhere. WS 3 cm. **Small Eggar** *Eriogaster lanestris* (2) is a very early spring moth, out in February and March, widespread but local and commonest in the S. The caterpillars make webs on Hawthorn and Blackthorn. WS 3 cm.

Lackey *Malacostroma neustria* (3). Locally common in the S of both England and Ireland; its caterpillars live in a web, feeding on Hawthorn and various other trees and shrubs. WS 3 cm. **Fox Moth** *M. rubi* (4) is found almost throughout Britain and Ireland. A fairly large moth, it has conspicuous hairy caterpillars that feed, often in numbers but not colonially, on Heather and other plants. WS 5 cm.

Emperor Moth *Saturnia pavonia*. A strikingly patterned moth that flies rapidly by day in April and May over heaths and moors in search of mates. Found almost throughout Great Britain and Ireland, it is again conspicuous in summer when the handsome, bright green caterpillars (p.218) can be seen on Heather, Bramble and a great variety of other plants. WS 5–6 cm.

Peach Blossom *Thyatira batis*. This very pretty moth is widespread and fairly frequent in most parts of Great Britain and Ireland, but rather local in Scotland. Since the favoured food-plant of the caterpillars is Bramble, it is surprising that the moth is not commoner. WS 3.5 cm.

Scalloped Hook-tip *Drepana lacertinaria* (5) has, like all our 5 hook-tips, a curious hooked tip to its forewings. It is widespread, found almost wherever there is Birch, on which the caterpillars feed, but commonest in the S. WS 2.5 cm. **Pebble Hook-tip** *D. falcataria* (6) is equally widespread, but more local in Ireland, and also feeds on Birch. WS 3 cm.

Lime Hawk-moth *Mimas tiliae*. Hawk-moths generally are large stout moths with large stout green caterpillars, usually with a horn at the rear end. The Lime Hawk-moth is one of the smaller ones, emerging in May or June. It is not uncommon in the S of England, but becomes less so northwards through the Midlands to the borders of Yorkshire. The caterpillar feeds on the foliage of lime and elm trees. WS 6 cm.

Pine Hawk-moth *Hyloicus pinastri*. A rather dull hawk-moth that nevertheless can often be picked out on the bark of pine trees, where it rests during the daytime between May and July. The caterpillar feeds on the needles of pine trees, especially the Scots Pine *Pinus sylvestris*. Fifty years ago the moth was rare, but is now more frequent where pines are found in S England. WS 8 cm.

Eyed Hawk-moth *Smerinthus ocellata*. The 'eyes' on its underwings make this one of the most striking of British hawk-moths. It is also one of the commoner ones, especially in the S half of England, but thinning out northwards and in Ireland. The caterpillars feed mainly on various kinds of willow and sallow, *Salix* spp., but also on Apple and a number of other trees and shrubs. WS 9 cm.

Poplar Hawk-moth *Laothoe populi*. Another striking hawk-moth, with its reddish-chestnut patches, but like the Pine Hawk-moth hard to detect when resting on a tree-trunk by day. The moths fly in May and June, and the caterpillars feed on various kinds of poplar, *Populus* spp., and willow, *Salix* spp. One of the commonest and most widespread British hawk-moths, it extends right up to the Scottish Highlands, except the extreme N. WS 7 cm.

Privet Hawk-moth *Sphinx ligustri*. One of the larger British hawk-moths, it is not uncommon in S England, thinning out northwards into S Scotland. It flies in June and July, when it lays the eggs that result in the handsome pink-striped caterpillars which feed on both Wild and Garden Privet, *Ligustrum vulgare* and *L. ovalifolium*. WS 9 cm.

Elephant Hawk-moth *Deilephila elpenor*. This is one of the smaller British hawk-moths, whose name derives from the supposed resemblance of the front part of the caterpillar to an elephant's trunk. The moth flies in June and is not uncommon in England and Wales, extending into S Scotland. The caterpillar feeds on Great Willow-herb *Epilobium hirsutum* and Marsh Bedstraw *Galium palustre*. WS 7 cm.

Hummingbird Hawk-moth *Macroglossum stellatarum*. A day-flying moth which is a frequent immigrant from the Mediterranean, it is often mistaken for a hummingbird as it hovers in front of flowers while extracting nectar from them with its long tongue. Caterpillars have been found on various species of bedstraw, but the moth does not seem to breed often in Britain. WS 4.5 cm.

Garden Carpet *Xanthorhoë fluctuata*. One of the commonest and most widespread of the large family of geometrid moths, characterised by their 'looper' caterpillars, which have only 2 pairs of claspers or false legs at the rear. They progress by stretching to their full length and then looping up to bring the claspers forward. This species flies throughout the spring and summer, the caterpillars feeding on plants of the Cabbage Family. WS 3 cm.

Magpie Moth *Abraxas grossulariata*. A common and strikingly coloured geometrid moth, widespread throughout the greater part of Great Britain and Ireland. Its caterpillars (p.218) are conspicuously coloured, indicating that they are distasteful to predators; they feed on currant and gooseberry bushes, and in gardens sometimes on the shrub *Euonymus japonicus*. The moth flies in July and August. WS 4 cm.

Yellow Shell *Euphyia bilineata*. A very common geometrid moth occurring almost throughout the British and Irish mainlands, in hedges, gardens and other bushy places, flying during most of the summer. The caterpillars feed on various low-growing weeds. WS 3 cm.

Large Emerald *Geometra papilionaria* (1). The emeralds are a group of attractive pale green moths, one of which, the **Essex Emerald** *Thetidia smaragdaria*, is now officially in danger of extinction. The Large Emerald is widespread and frequent throughout Britain and Ireland, except for the N parts of Scotland. It flies in June and July, and the caterpillars feed on Birch, Beech and Hazel. WS 4.5 cm.

Peppered Moth *Biston betularia*. This moth was used by the late Dr Bernard Kettlewell and others to demonstrate the evolution of industrial melanism – the dark form of the moth survives better in smoky districts, where its colour camouflages it against predation by tits and other birds. It is widespread and locally common, flying in May and June. The caterpillar is shown on p.218. WS 4 cm.

Blood-vein *Calothysanis amata*. A strikingly patterned moth that is widespread and locally frequent in England, but much more local in Scotland, Ireland and Wales. It flies in June, and sometimes again in August. The 'looper' caterpillar feeds on various low-growing weeds. WS 3 cm.

Winter Moth *Operoptera brumata*. A winter-flying geometrid moth, occurring from October to February, whose female is wingless. The sticky bands often seen round the trunks of fruit trees are placed there to stop this wingless female, among others, from climbing up and laying her eggs on the twigs above. The 'looper' caterpillars (shown on p.218) feed on a wide variety of trees and shrubs. WS 2.5 cm.

Mottled Umber *Erannis defoliaria*. Another winter-flying moth, with much the same life history as the Winter Moth (above), and against whose wingless females fruit growers also have to defend their trees by using grease bands. Both the Winter Moth and the Mottled Umber are common, but the latter thins out northwards in Scotland. WS 4 cm.

August Thorn *Ennomos quercinaria*. The thorn moths all have the same slightly angular wing shape. This species flies in August and September and is widespread in England, though much commoner in the S, and in Ireland, but rather rare in Scotland. The caterpillar feeds on various trees and Hawthorn. WS 4 cm.

Brimstone Moth *Opisthograptis luteolata*. A widespread and common yellow moth, which can be found abroad in most of the spring and summer months, but seems to be most frequent in May or June. The 'looper' caterpillars feed mainly on Hawthorn, but can also be found feeding on Plum and Blackthorn. WS 3.5 cm.

Swallowtailed Moth *Ourapteryx sambucaria*. A widespread and common moth that is much rarer in Scotland than elsewhere in Britain and Ireland. It flies in July, and its 'looper' caterpillars feed on Ivy, Elder, Hawthorn, Blackthorn and other shrubs. They are twig-like, and therefore well camouflaged. WS 5 cm.

217

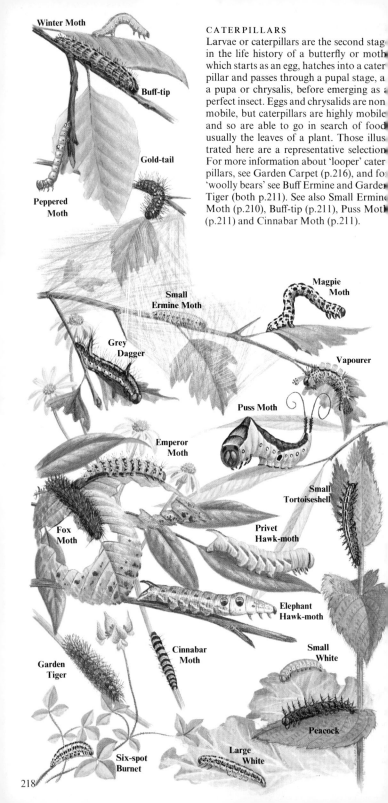

CATERPILLARS

Larvae or caterpillars are the second stage in the life history of a butterfly or moth which starts as an egg, hatches into a caterpillar and passes through a pupal stage, as a pupa or chrysalis, before emerging as a perfect insect. Eggs and chrysalids are non-mobile, but caterpillars are highly mobile and so are able to go in search of food, usually the leaves of a plant. Those illustrated here are a representative selection. For more information about 'looper' caterpillars, see Garden Carpet (p.216), and for 'woolly bears' see Buff Ermine and Garden Tiger (both p.211). See also Small Ermine Moth (p.210), Buff-tip (p.211), Puss Moth (p.211) and Cinnabar Moth (p.211).

Winter Moth

Buff-tip

Gold-tail

Peppered Moth

Small Ermine Moth

Magpie Moth

Grey Dagger

Vapourer

Puss Moth

Emperor Moth

Small Tortoiseshell

Fox Moth

Privet Hawk-moth

Elephant Hawk-moth

Cinnabar Moth

Small White

Garden Tiger

Peacock

Six-spot Burnet

Large White

Other Invertebrates

The invertebrates comprise the whole animal kingdom, except for a large group of phylum called the Chordata, which contains the vertebrates (pp 135–198). The phylum Arthropoda, which is in fact the largest animal phylum, containing more than 85 per cent of all the world's animal species, covers the great majority of species described and illustrated in this section. There are 12 classes of arthropods, of which the most important are the Diplopoda (millipedes), Chilopoda (centipedes), Insecta (insects), Crustacea (crustaceans) and Arachnida (spiders and mites).

The largest and best-known of the 27 classes of our insects are the Odonata (dragonflies), Orthoptera (grasshoppers), Phthiraptera (lice), Hemiptera (bugs and aphids), Coleoptera (beetles), Lepidoptera (butterflies and moths; see p.199), Diptera (2-winged flies), and Hymenoptera (bees, wasps and ants).

The **Arachnida** are distinguished from the insects by having 4 pairs of walking legs, and only the spiders have their body obviously divided into 2 parts. Over 1,500 species are known in Britain and Ireland. The main groups are the Araneae (spiders), Opiliones (harvestmen), Pseudoscorpiones (false scorpions) and Acari (mites and ticks).

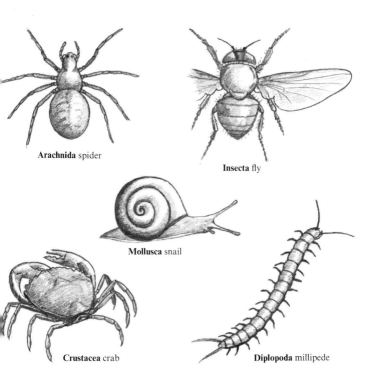

Arachnida spider

Insecta fly

Mollusca snail

Crustacea crab

Diplopoda millipede

The **Insecta** are the most abundant class of arthropods on land; approximately 21,500 species are known to occur in Britain and Ireland. The bodies of adult insects are divided into head, thorax and abdomen, the head bearing a pair of feelers or antennae, and the thorax 3 pairs of legs and usually also wings. Insects are in fact the only winged arthropods. Their life history usually passes through the 3 stages of egg, larva and pupa before they attain adulthood.

The **Crustacea** are mainly a marine group, characterised by their 2 pairs of antennae, but some species occur in fresh water, and one group, the Isopoda (woodlice) are terrestrial.

The **Mollusca** are the second largest phylum of higher animals, but are again mainly marine, though some snails and slugs and bivalves are found on fresh water and on land. Molluscs have a (usually) hard shell and a tough muscular organ known as the foot.

BRISTLE-TAILS Thysanura. A primitive wingless insect group. **Silverfish** *Lepisma saccharina* (1) is named from its bright, silvery scales. Common in kitchens and libraries, it often eats the glue in book bindings. 1 cm.

SPRINGTAILS Collembola. This is a group of tiny, wingless insects, rarely as much as 5 mm long, found in vast numbers in the soil. The species shown (2) is *Podura aquatica*, one of our 2 freshwater springtails, found at the edge of ponds.

MAYFLIES Ephemeroptera. A group of aquatic insects much eaten by fish. After hatching, mayflies become first nymphs, then duns, and finally spinners; nymphs are aquatic, the other 2 aerial in habit. Mayflies appear in May and can be found throughout the summer. Those shown are the **Greendrake** *Ephemera danica* (3), L (less tail) 2 cm, WS 4.5 cm, and the **Pond Olive** *Cloëon dipterum* (4), L (less tail) 5 mm, WS 1.5 cm.

♂ **Common Sympetrum Dragonfly**

♂

Banded Agrion Damselfly

Emperor Dragonfly

220

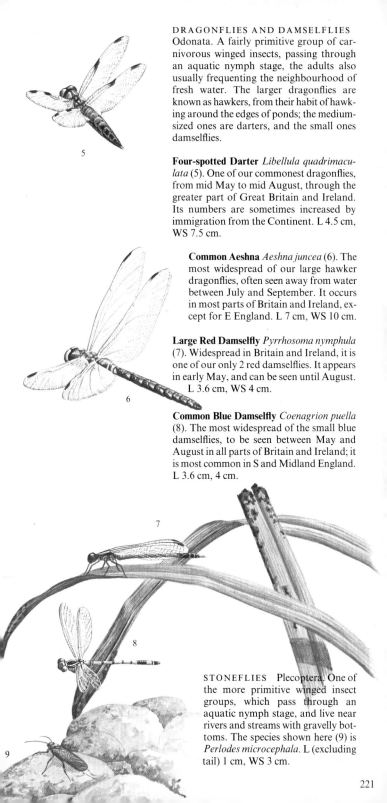

DRAGONFLIES AND DAMSELFLIES

Odonata. A fairly primitive group of carnivorous winged insects, passing through an aquatic nymph stage, the adults also usually frequenting the neighbourhood of fresh water. The larger dragonflies are known as hawkers, from their habit of hawking around the edges of ponds; the medium-sized ones are darters, and the small ones damselflies.

Four-spotted Darter *Libellula quadrimaculata* (5). One of our commonest dragonflies, from mid May to mid August, through the greater part of Great Britain and Ireland. Its numbers are sometimes increased by immigration from the Continent. L 4.5 cm, WS 7.5 cm.

Common Aeshna *Aeshna juncea* (6). The most widespread of our large hawker dragonflies, often seen away from water between July and September. It occurs in most parts of Britain and Ireland, except for E England. L 7 cm, WS 10 cm.

Large Red Damselfly *Pyrrhosoma nymphula* (7). Widespread in Britain and Ireland, it is one of our only 2 red damselflies. It appears in early May, and can be seen until August. L 3.6 cm, WS 4 cm.

Common Blue Damselfly *Coenagrion puella* (8). The most widespread of the small blue damselflies, to be seen between May and August in all parts of Britain and Ireland; it is most common in S and Midland England. L 3.6 cm, 4 cm.

STONEFLIES Plecoptera. One of the more primitive winged insect groups, which pass through an aquatic nymph stage, and live near rivers and streams with gravelly bottoms. The species shown here (9) is *Perlodes microcephala*. L (excluding tail) 1 cm, WS 3 cm.

221

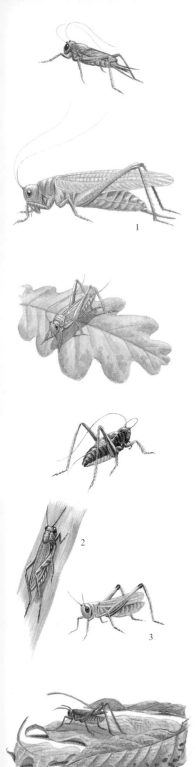

GRASSHOPPERS AND CRICKETS
Orthoptera

House Cricket *Acheta domesticus*. An alien relative of the grasshoppers, which needs a warm climate, and so can only live in kitchens, bakeries, rubbish dumps and other warm places, where its incessant chirruping can often be heard. It is widespread and frequent in England, but more local in the cooler climate of Scotland, Wales and Ireland. 1.5 cm.

Great Green Bush Cricket or **Great Green Grasshopper** *Tettigonia viridissima* (1). Our largest bush cricket, which is not infrequent in parts of S and Midland England and S Wales. It is most often found among bushes, hedges and coarse vegetation, and utters the loudest song of any of our Orthoptera, except the rare **Field Cricket** *Gryllus campestris*. Former species 4–5 cm.

Oak Bush Cricket *Meconema thalassinum*. The only British orthopteron that lives in trees, especially oaks, it is widespread and locally frequent in S and Midland England and Wales, much less so in N England and in Ireland, and is absent from Scotland. Primarily nocturnal in habit, it does not begin drumming till after dark. 1.2 cm.

Dark Bush Cricket *Pholidoptera griseoaptera*. Fairly common in the S half of England and S Wales, it does not extend N of Yorkshire, and is altogether absent from Scotland and Ireland. It is catholic in its choice of habitat, almost wherever there are bushes or coarse vegetation, and is an early starter, often appearing in April. Its 'song' is a short, sharp chirp. 1.5 cm.

Common Field Grasshopper *Chorthippus brunneus* (2) and **Meadow Grasshopper** *C. parallelus* (3) are 2 of our commonest grasshoppers of grassy places. Both are very variable in colour, in differing shades of green and brown, and occur throughout the mainland of Great Britain. However, the first is much less common in the Scottish Highlands but present in Ireland, while the second is commoner in Scotland but absent from Ireland. Both species 1.5–2 cm.

Common Ground-hopper *Tetrix undulata*. A small, grasshopper-like insect found in most parts of Britain and Ireland, but rather local except in England. It likes dry conditions, especially in woods and on heaths, but is shy and not often seen. Nor can it be heard, since it does not stridulate. 1 cm.

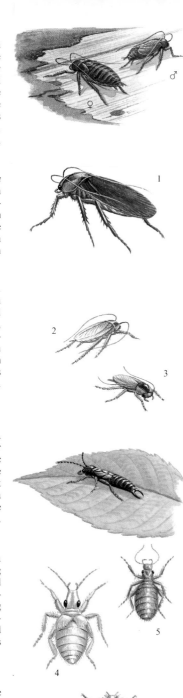

COCKROACHES Dictyoptera
Common Cockroach *Blatta orientalis*. An alien insect with a flattened appearance that needs a warm climate, and so is only found indoors, scavenging on scraps of food left uncovered. Despite their popular name of 'black beetle', Common Cockroaches are only superficially beetle-like. Their wings are scarcely developed, so they cannot fly. They are much less common in N Scotland, Ireland and Wales. 2 cm.

American Cockroach *Periplaneta americana* (1). A large, reddish brown cockroach with fully developed wings. 3 cm. Its equally unwelcome relative the **Australian Cockroach** *P. australasiae* can be distinguished by the pale yellow stripes on its forewings. Both are much less common than the Common Cockroach.

German Cockroach *Blattella germanica* (2). Smaller and less common than the Common Cockroach, this species has fully developed wings. It probably originates from N Africa, as reflected in its liking for bakeries, hot-houses and similar places with a warm microclimate. 1.2 cm. **Dusky Cockroach** *Ectobius lapponicus* (3) is a native species found mainly on the ground in S England. 1.5 cm.

EARWIGS Dermaptera
Common Earwig *Forficula auricularia*. A widespread, common and familiar member of the Dermaptera, with a characteristic small pair of forceps at the hind end of the body. It likes skulking in dark places, but there is no real evidence that it ever enters a human ear, except quite by chance. The young pass through a nymphal stage. 1.5 cm.

BOOK-LICE Psocoptera
These tiny insects fall into 2 main groups: a large one of mainly winged species living on the bark of trees or in undergrowth, and called bark-lice, and a smaller one of wingless species inhabiting buildings and feeding on moulds and starchy materials, the book-lice proper. Illustrated here are the winged *Atropus pulsatorius* (4) and the wingless *Liposcelis divinatorius* (5). Both 1–2 mm.

FEATHER LICE Mallophaga
A group of small, wingless insects that live on the skin, mainly of birds but also of a few mammals, and feed on tiny fragments of skin, hair and feathers. Their complete life history takes place on the body of their host. Illustrated here is a species of *Culutogaster*. 1–2 mm.

SUCKING LICE Anoplura

Members of the Anoplura group are found only on mammals, whose blood they suck, as distinct from the feather lice or Mallophaga (p.223), which live mainly on birds. The **Human Louse** *Pediculus humanus*, illustrated here, exists in 2 forms, the Head Louse *P. h. capitis* and the Body Louse *P. h. corporis*. 2 mm.

eggs on human hair

BUGS Hemiptera

Bed Bug *Cimex lectularius*. Formerly a well-known blood-sucking pest of dirty houses, lurking in cracks in the woodwork and emerging at night, it is now quite scarce in Britain and Ireland, thanks to DDT. Its subspecies *columbarius* preys on pigeons in dovecotes. In heated rooms breeding is said to go on throughout the year. 4 mm.

Hawthorn Shield Bug *Acanthosoma haemorrhoidale*. A plant bug, common on Hawthorn in England, Wales and Ireland, especially in woodland. It feeds on Hawthorn berries and leaves, and can be found in town parks and such urban open spaces as Hampstead Heath in London. 1 cm.

Pied Shieldbug *Sehirus bicolor*. A common plant bug of waysides and waste ground, where its favourite food-plant, White Deadnettle *Lamium album* (p.77), grows. It may also feed on Black Horehound *Ballota nigra* (also p.77). The adults live inside the flowers and their larvae feed on the plant's nutlets. It is common in England and Wales S of Lancashire and Yorkshire, but is absent from Scotland and Ireland. 6 mm.

Firebug *Pyrrhocorus apterus*. This strikingly coloured plant bug is widespread on the Continent, but in Britain, apart from casual immigrants, it has only one permanent colony, on a small island off the S coast of Devon, where it has been known for more than 170 years. In Europe it congregates in masses on Mallows and Limes. 1 cm.

Common Flowerbug *Anthocoris nemorum*. One of our commonest bugs, widespread throughout the mainlands of Great Britain and Ireland. A predator, it feeds on a wide variety of tiny insects, such as the Red Spider Mite and greenfly, and its incidental destruction by the careless and indiscriminate use of insecticides by gardeners is one cause of plagues of these pests. 5 mm.

Water Measurer or **Water Gnat** *Hydrometra stagnorum*. An elongated, wingless water bug, 1–1.5 cm long, which lives entirely in the surface film of still fresh water, and among the vegetation on the edge of ponds. It is enabled to do this by the fine hairs that line its under-surface and prevent its body from becoming wet. 1 cm.

Pond Skater or **Water Strider** *Gerris lacustris*. A very common aquatic bug, larger than the Water Measurer, which also skims the surface film of lakes and ponds, never breaking the surface to get wet. It is a scavenger, feeding on various luckless insects that fall into the water and drown. 8 mm.

Water Boatman, Wherryman, Backswimmer or **Boat-fly** *Notonecta glauca*. A common predatory aquatic bug that swims upside-down just under the surface of the water, propelling itself by its powerful hind-legs. It is very active and aggressive, attacking creatures larger than itself, even fish, and is reputed to be able to prick human skin if given the chance. 1 cm.

Water Stick Insect *Ranatra linearis*. A remarkably long, thin aquatic bug, some 5 cm in length; its breathing tube occupies nearly half its length. It is mainly southern in distribution. The Water Stick Insect is not at all closely related to the terrestrial stick insects, which are not native to Britain.

Water Scorpion *Nepa cinerea*. A widespread and common elongated water bug, superficially resembling a terrestrial scorpion (which is actually not even an insect, but related to the spiders) on account of its long, tail-like breathing tube, the supposed 'sting' of the scorpion. It is a predator on small water creatures. 2 cm.

Saucer Bug *Ilyocoris cimicoides*. A squat, sluggish water bug found at the muddy margins of stagnant waters, mainly in the S. It is a predator, and another insect that can pierce human skin if handled carelessly. 1 cm.

spit

colour
variations

Cuckoo Spit Insect *Philaenus spumarius*. Cuckoo spit is the foam produced by the nymphs of the bugs which are also called froghoppers because they jump, probably to protect themselves against desiccation. The froth is produced by discharging air through a valve in the abdomen to mix with liquid issuing from the anus, and is often abundant on plants in summer. 8 mm.

Red and Black Froghopper *Cercopis vulnerata*. One of the most strikingly coloured of the froghoppers or cuckoo spit insects, which are members of the Homoptera division of the bugs (Hemiptera), unlike the plant bugs, which are in the Heteroptera division. The Red and Black Froghopper is quite common, and its nymphs, like those of other froghoppers, suck plant juices. 1 cm.

Cabbage Whitefly *Aleyrodes brassicae* (1). Whiteflies are a family (Aleyrodidae) of small bugs, whose wings and bodies are coated with a mealy white powder and so resemble minute moths or midges. They exude honeydew. Some whiteflies are economic pests, such as the Cabbage Whitefly and the **Greenhouse Whitefly** *Trialeurodes vaporariorum*, which infests potted tomatoes. Both 1–2 mm.

Woolly Aphid or **American Blight** *Eriosoma lanigerum*. Aphids are small homopteran plant bugs, often green, that suck plant juices and tend to congregate in large numbers on both wild and garden plants. Woolly Aphid is an alien, common in neglected apple orchards, which produces a waxy substance that looks like cotton wool. 2 mm.

Black Bean Aphid *Aphis fabae*. One of the best-known aphid pests, often abundant on broad beans, which overwinters in the egg on the Spindle tree. Ladybird larvae are well-known predators on aphids, and care must therefore be taken when spraying aphids in the garden not to kill their predators too, or next year's plague will be worse than this year's. 2 mm.

Peach and Potato Aphid *Myzus persicae* (2). Yet another garden pest, whose tastes are clear from its English name. Perhaps the best known of all aphids is the **Greenfly** *Siphocoryne rosarum*, which infests garden roses. Some species of aphid are milked by ants for their honeydew. 2 mm.

THRIPS Thysanoptera
A group of small, sap-sucking flies, sometimes known as Thunder Flies, with characteristic fringes of long hairs on their wings. The tiny flies so common among the florets of Dandelions and other composites are Thrips. Some species are pests, e.g. of peas and onions. The name 'Thrips' is both singular and plural. 2mm.

LACEWINGS Neuroptera
Green Lacewing *Chrysopa carnea* (1). Insects with weak flight, which frequent foliage during the summer, but in the autumn often enter houses to spend the winter in hibernation, turning red until they emerge, green once again, in the spring. Their larvae feed on Greenfly (2), other aphids and plant bugs, also on mites. WS 2.5 cm.

Giant Lacewing *Osmylus fulvicephalus*. One of the largest of the lacewings, this species is semi-aquatic, the larvae living in wet moss at the edge of streams, where they feed by sucking the blood of smaller larvae, such as those of the chironomid midges, known as Bloodworms, which are also found living in the moss. WS 4.5 cm.

SCORPION FLY Mecoptera
A small group of weakly flying, predatory winged insects whose name is derived from the male's habit of carrying the tip of its body curled up over its back, in the manner of a scorpion's tail. The species illustrated is the common *Panorpa communis*, often found in hedges, which may visit ripe fruit in gardens. WS 2.5 cm.

CADDIS FLIES Trichoptera
Also known as sedge-flies, which are aquatic insects beloved of the fly-fisherman, who imitates them with his artificial flies to catch the trout who also love them. The **Great Red Sedge** or **Murragh** *Phryganea grandis* is our largest caddis, up to 2 cm in length. WS 6.5 cm.

Grannom or **Green-tail** *Brachycentrus subnubilus*. One of the first caddis flies to be seen in the spring, this species often appears as early as the month of April in the S of England, for instance on the River Kennet, the well-known trout stream in Berkshire. L 5 mm; WS 1.8–2 cm.

TWO-WINGED FLIES Diptera

Daddy Long Legs or **Crane Fly** *Tipula paludosa*. One of the commonest of the Tipulidae or Diptera, whose larvae, called leather-jackets and abundant in grassland, can be a serious agricultural pest. They feed on the roots and lower stems of a wide variety of plants, and are relished by Rooks, among other birds. 3 cm.

Common Gnat *Culex pipiens*. Our commonest mosquito, often found in houses, but fortunately it does not bite people, only birds. Its larvae can be found in stagnant water, such as neglected water butts, wells and puddles. The larvae are very active, moving about in the water by jerks, doubling up their bodies and straightening them out again. 6 mm.

Ringed Mosquito *Theobaldia annulata*. This is the villain that does bite humans, and for whose misdeeds the harmless Common Gnat often gets swatted. The Ringed Mosquito can be distinguished by the dark spots on its wings. Its life history is very similar to that of the Common Gnat, including hibernation indoors and emerging on warm winter days. 9 mm.

Blackfly *Simulium equinum*. These are the small, black, 2-winged flies, related to the gnats and midges, which cause great annoyance by swarming round people's heads and sometimes biting them. They must not be confused with the Blackfly that is a pest on broad beans, which are aphids (p.226). The larvae of *Simulium equinum* are aquatic. 4 mm.

larvae

Midge *Chironomus plumosus*. One of the swarming, non-biting chironomid midges, whose tiny, bright red larvae, found in water butts and stagnant pools, are called blood-worms. The adults dance up and down in sheltered places near water. The larvae progress by making figures-of-eight. 8 mm.

Horse Fly *Tabanus bovinus*. One of the commonest of the blood-sucking tabanid 2-winged flies. Unlike Clegs (p.229), which are silent, tabanids often give warning of their approach by a high-pitched humming noise. When such a large fly bites, it may make a hole large enough for a drop of blood to ooze out. In the New Forest tabanids are called stouts. 1.2 cm.

228

Cleg *Haematopoda pluvialis* (1). Tiresome blood-sucking flies that go under many local names, such as burrel flies, whame flies, breeze flies and dun flies. This species is commoner in the S; in the N and in Scotland the closely related *H. crassicornis* is more frequent. The silent approach of these flies and their sharp, stabbing bite seem to be more often experienced in close, thundery weather. 1 cm.

Bee Fly *Bombilius major*. A completely harmless 2-winged fly that bears a remarkable resemblance to a small bumble-bee. It flies quite early in the year, in April and May, and often hovers in the air with its wings vibrating so fast that they cannot be seen. The larvae parasitise solitary bees. 1.2 cm.

Robber Fly or **Assassin Fly** *Asilus crabroniformis*. A large, predatory, 2-winged fly that owes its Latin name to its likeness to a small Hornet. It attacks other flying insects, sometimes larger than itself, and sucks out their juices. Tiny flies of the genus *Desmometopa* sometimes hitch a lift on its body and are thus enabled to participate in the feast. 2 cm.

Hover Fly *Volucella bombylans* (2). One of the bigger species in the large group of hover flies, noted for their habit of hovering apparently motionless in the air and darting away swiftly out of sight. Some of their larvae are known as red-tailed maggots. 1.4 cm. Some adults, including the genus *Volucella*, are remarkably like bees. *Syrphus* (3) is another important genus of hover flies. 8–10 mm.

Drone Fly *Eristalis tenax*. A large hover fly that quite closely resembles a drone Honeybee. Its larva is one of the rat-tailed maggot type, and is often found in the old-fashioned kind of farmyard where the mud gets fouled with animal dung. When caught in an entomologist's net or a spider's web, the males emit a very high-pitched note. 1.5 cm.

Horse Bot Fly *Gasterophilus intestinalis*. A 2-winged fly related to the common House Fly, whose larvae infest horses, donkeys and mules. The eggs are usually laid on the animal's fore-legs, and cause an irritation which makes the horse lick them up and swallow them, from where they make their way to the lining of its stomach. Their presence can make horses panic. 1.5 cm.

Vinegar Fly or **Fruit Fly** *Drosophila melanogaster*. One of a large group of minute 2-winged flies whose larvae's salivary glands have large chromosomes which make them particularly suitable for genetic research work. The flies are often found where fruit is stored or where vinegar or other fermented products such as wine, for example, are being made. 3 mm.

Frit Fly or **Oat Fly** *Oscinella frit*. One of the best-known agricultural pests, which can do serious damage to cereals, particularly oats. Only 1–1.5 mm long, it first appears in April, and produces 3 generations a year during the growing season of the grasses and cereal crops on whose stems and grains the larvae feed.

Warble Fly *Hypoderma bovis*. These, and not the Clegs or Horse Flies, are the true gadflies, whose presence and bites make cattle 'gad' about with their tails erect. Their larvae feed just under the skin of cattle, producing a swelling which can ruin the hide; the Warble Fly is therefore one of the more serious economic pests in agriculture. 3 cm.

Bluebottle or **Blowfly** *Calliphora erythrocephala* (1). This is the well-known large, metallic blue 2-winged fly, which is all too familiar in the house, whose larvae feed on decaying animal matter and dung, and on any uncovered meat. 1.2 cm. The closely related *C. vomitoria*, with a red head, is an equal pest.

Greenbottle *Lucilia caesar*. A metallic green 2-winged fly, whose larvae are among the maggots seen in decaying meat and other animal substances. 1 cm. *L. bufonivora* has the singularly unpleasant habit of laying its eggs on the eye-rims and nostrils of frogs and toads, usually leading to the death of the host. *L. sericata* is the cause of the complaint known as 'strike' in sheep.

Flesh Fly *Sarcophaga carnaria*. This is another 2-winged fly whose larvae (maggots) feed on carrion, or on meat intended for human consumption. Fortunately they rarely come indoors, where they would be an unhygienic pest. The young are born as maggots, not as eggs. 1.6 cm.

Cluster Fly *Pollenia rudis*. A relative of the Greenbottles (opposite), named from its habit of clustering together in crevices or corners of buildings in late autumn. In the summer it basks in the sun on tree-trunks and other bare surfaces. The larvae are parasitic on earthworms. 7 mm.

Yellow Dung Fly *Scatophaga stercoraria*. These are the repulsive-looking flies which assemble on cowpats, the males awaiting the arrival of the females who, after mating, will lay their eggs in the dung. They are predatory and will attack and kill other flies feeding on the nectar of various sweet-scented flowers. 1 cm.

House Fly *Musca domestica* (1). An all-too-common pest of houses, adept at conveying bacteria from the unsavoury refuse where it lays its eggs to the human food, especially sugar, on which it feeds. 8 mm. Smaller individuals are not young House Flies but the **Lesser House Fly** *Fannia canicularis*. 6 mm. An orange–yellow, 2-winged fly in the house is likely to be the **Biting House Fly** *Musca autumnalis*.

Sheep Ked, Sheep Louse or **Sheep Tick** *Melophagus ovinus*. This is actually a wing-less, 2-winged fly; in other words it belongs to the Diptera, like the House Fly. In fact it looks more like a spider as it crawls about the fleeces of sheep. By causing irritation it makes sheep rub their fleece off in patches, creating an opportunity for other pests, such as Blow-flies. 6 mm.

FLEAS Siphonaptera
Fleas are found on many species of both bird and mammal (e.g. **Dog Flea** *Ctenocephalides canis* (2), 2–3 mm, the largest British flea being the **Mole Flea** *Hystricopsylla talpae*. The **Rabbit Flea** *Spilopsyllus cuniculi* is the chief carrier of the well-known disease of Rabbits, myxomatosis. **Human Flea** *Pulex irritans* (3) is, like the Louse and the Bed-bug, less familiar nowadays than it used to be. 2–4 mm.

BEES, WASPS AND ANTS
Hymenoptera
Willow Sawfly *Pontania proxima*. This species is the one which produces the bright red bean galls on willow leaves. These galls, which grow in response to some chemical stimulus applied by the female Sawfly, each house first an egg and then a larva, which eventually goes off to pupate elsewhere. 8 mm.

Giant Wood Wasp or **Greater Horntail** *Urocelus gigas*. One of our largest wasps, liable to be mistaken for a Hornet. Its apparent sting, however, is a very long ovipositor, which it buries deep in rotten wood when laying its eggs. The rest of its life history, larva and pupa, is spent in the depths of the log, where the larvae are sought out by the Large Ichneumon (below). 3 cm.

Marble Gall Wasp *Andricus kollari*. A small wasp which produces one of the best-known plant galls, the marble gall of oak trees, also known as bullet gall and oak nut. Each gall contains a single larva. Marble Gall Wasps are not native to Britain, but were introduced from the Middle East in the 18th century because of the galls' high content of tannic acid, used in the tanning industry and in the preparation of ink. 8 mm.

Robin's Pincushion Gall Wasp *Rhodites rosae*. Another small wasp, it produces on wild roses the conspicuous mossy red and green galls known as robin's pincushions, bedeguars or moss galls. Several other species of tiny wasp take advantage of the situation to eat either the gall tissue or each other. 5 mm.

Oak Apple Gall Wasp *Biorhiza pallida*. Oak apples are among the best-known plant galls and named from their rosy pink appearance when mature in May and June. Afterwards they turn brown, and the fully grown wasps emerge through several holes. Oak Apple Day, 29 May, was named to commemorate the Restoration in 1660 of King Charles II, who hid in an oak tree after the Battle of Worcester. 4 mm.

Yellow Ichneumon *Netelia testacea*. Ichneumons are parasitic wasps which lay their eggs in the eggs or larvae of other insects. Their own larvae live inside their host, which dies when the ichneumon larvae emerge to pupate. This large species often comes to light in houses. It parasitises some of the larger moth larvae. 2 cm.

Large Ichneumon *Rhyssa persuasoria*. This is the largest British ichneumon, which parasitises one of the largest British insects, the Giant Wood Wasp (above). It uses its immensely long ovipositor to lay its eggs in the larvae of that wasp, which are deep inside rotten logs. 3 cm without ovipositor, 6·5 cm with ovipositor.

Cabbage White Ichneumon *Apanteles glomeratus*. A small parasitic wasp, best identified by the sight of the cluster of small yellow cocoons around the corpse of a Cabbage White butterfly larva, inside which the ichneumon larvae will have been living. 4–5 mm.

larvae

Common Red Ant *Myrmica rubra*. A common ant of lawns and grassland, this species is one of the group called dairying ants because they farm aphids and milk them for their honeydew. It also preys on other insects, including other ants. This is the ant in whose anthills the caterpillars of the Large Blue butterfly (p.208), which has recently become extinct in Britain, spend part of their lives. 3–6 mm.

Black Lawn Ant *Acanthomyops niger*. Another dairying ant of lawns and grassy places. As well as milking aphids, they derive a sticky substance from the caterpillars of various blue butterflies, which secrete it from special glands. The Black Lawn Ant is a very common ant near houses. Female 8–9 mm, male 3–5 mm.

Yellow Hill Ant *Lasius flavus* (1). A small ant, often found on chalk downs and in hill districts. Lost travellers are said to be able to re-orient themselves by the fact that the tall anthills of the Yellow Hill Ant are sited with their ends pointing E and W. Female 7–9 mm, male 3–4 mm.

1

nest

Wood Ant *Formica rufa* (2). This is one of the best-known ants, because it draws attention to itself in drier woodlands, especially pinewoods, by its large nests, often consisting of piles of pine needles. Often, too, the long trails of ants going to and from the nest can be observed. When the nest is threatened, the workers defend it by discharging a vapour of formic acid. 6–10 mm.

2

Pharaoh's Ant *Monomorium pharaonis*. This is the tiny, reddish yellow ant, a native of warmer climates, which infests buildings, especially those which are kept warm by central heating. It can be a serious pest. 2 mm.

233

Hornet *Vespa crabro.* Our largest social wasp, more than 2.5 cm long, which nests in hollow trees and sometimes also in the roof of a house or a hollow in a bank. The nest is made of a papery substance derived from decayed wood. Despite its fearsome reputation, the Hornet will sting man only if attacked. It is found mainly in the S.

Common Wasp *Vespa vulgaris* (1). Together with the very similar **German Wasp** *V. germanica* it is a most unwelcome visitor to kitchens and to meals eaten in the open air, such as picnics. The Common Wasp makes its papery nest in holes either in the ground or in trees. The somewhat similar **Tree Wasps** *V. sylvestris* and *V. norvegica,* on the other hand, hang their nests, which resemble those of the Common Wasp, from the boughs of trees or shrubs. All 1.2–1.5 cm.

Heath Potter Wasp *Eumenes coarctata.* A small wasp named from the flask-like nests the female constructs with tiny clay pellets, cemented with saliva, often on the branches of heaths and heathers, but occasionally on a post or tree stump. Each nest is provisioned with tiny larvae to feed the wasp's own larva when it hatches. 1.1–1.4 cm.

Spiny Mason Wasp *Odynerus spinipes.* Another small wasp which has remarkable nesting habits. The female burrows into a sandy bank, or occasionally into a soft-mortared wall, and constructs a small, curved edifice over the mouth of the burrow with the grains of sand she brings out. The burrow is then stored with small larvae of various kinds, to provision the eventual grub. 1.2 cm.

Red-banded Sand Wasp *Ammophila sabulosa.* This is another wasp which burrows in the sand to make a safe nest for its larva, in this case often in sand-dunes above the high-water mark. This species, however, captures large caterpillars, bigger than itself, which it paralyses and then drags to its burrow, sometimes over a distance of several metres. 1.2–2 cm.

Ruby-tail Wasp *Chrysis viridula.* A most handsome small red and green wasp, which preys on the Mason Wasps (above). It enters their burrows and lays its own egg in the well-stocked larder. The resulting Ruby-tail grub eats not only the larva of its Mason Wasp host, but also all the food supply so carefully amassed by the female Mason Wasp. 6–9 mm.

Mining Bee *Andrena armata*. One of a large group of small bees that excavate burrows in sandy soil, where they lay their eggs in cells provisioned with pollen and honey for their grubs to eat. They are preyed on by another group of bees, called *Nomada*, which seek out the burrows of the *Andrena* bees and then lay their own eggs in them. 1.2 cm.

Patchwork Leaf-cutter Bee *Megachile centuncularis*. These small bees belong to the larger group of carpenter bees, and construct their nest cells from pieces cut from leaves. Some half-dozen cells are built in a pile, one on top of the other, each containing a single egg. This is the species which cuts semi-circular pieces out of the leaves of garden roses. 8–12 mm.

Buff-tailed Bumblebee *Bombus terrestris*. One of our commonest bumblebees, which builds its nest underground, sometimes using an old burrow, possibly a metre or so long, made by a mouse or vole. The female bee collects for her grubs honey which she stores in a special honeypot, constructed of the same wax, secreted by her body, as the cells. 1.5–2.2 cm.

Large Red-tailed Bumblebee *Bombus lapidarius*. The largest of several species of bumblebee that have a red, not a yellow, tip to the abdomen. Common in England, it thins out northwards in Scotland. Its life history is similar to that of the Buff-tailed Bumblebee (above), but it will sometimes build its nest in the chinks of an old wall or the banks of a West Country stone hedge. 1.5–2 cm.

Common Carder Bee *Bombus agrorum* (1). Similar to the Buff-tailed Bumblebee, but smaller; the rear end of the abdomen may be dull brown rather than yellow. It builds its nest on or above the ground. Of about the same size is the **Small Earth Bumblebee** *B. lucorum*, whose tail-tip is usually whitish. It may nest either above or below ground. 1.2–2 cm.

Honeybee *Apis mellifera*. The most important honey-gathering social bee, and the only one which it has been worthwhile for man to induce to nest in specially constructed beehives. Far more Honeybees now nest in such hives than in natural situations in the wild. The characteristic original British form of the Honeybee is dark brownish black. 1.2–1.8 cm.

Green Tiger Beetle *Cicindela campestris* (1). A common predatory beetle of sandy places in spring and early summer, it received its name because of its ferocity in attacking its prey. In colour it may range from bright green to almost black. 1.5 cm. Its relative the **Wood Tiger Beetle** *C. sylvatica*, also found in sandy places, is a purplish bronze colour. 1.5 cm.

Violet Ground Beetle *Carabus violaceus* (2). A very common species of 'black beetle', it is named from the colour of the edges of its elytra, or hardened forewings, which may be violet, purple, blue or red. 2.5 cm. Another species, whose whole elytra may be reddish, is *C. granulatus*, which is found in damp places.

Whirligig Beetle *Gyrinus natator*. This species is one of a family of small beetles noted for their habit of gyrating around one another on the surface of still or slow-moving water, sometimes diving below the surface if danger appears to threaten them. The beetles lay their eggs on plants that are submerged below the surface of the water. 5–6 mm.

Great Water Beetle *Dytiscus marginalis*. One of our largest beetles, this species spends the majority of its time under water, but it still has to swim up to the surface of the water from time to time in order to take in a supply of air through its hindquarters. Like most water beetles it is predatory and feeds on smaller aquatic creatures. 2.6–3.2 cm.

Great Silver Beetle *Hydrophilus piceus*. This species is a water beetle that is larger even than the Great Water Beetle, but differing in its purely vegetarian habits, though its larva is predatory. This species differs also in the fact that it is all black, without any yellow facings on its upper surface. 3.7–4.8 cm.

Burying Beetle *Necrophorus vespilloides* (3). This species is one of a group of carrion beetles which are also known as sexton beetles from their habit of burying the carcases of small animals, such as mice, and rolling up the skinned flesh of these creatures into balls to provide food for their larvae. 1–1.5 cm. The often commoner species *N. humator* has the rear of its 2 orange bands rather more clearly defined than it is in *N. vespilloides*.

Devil's Coach Horse, Cock Tail or **Rove Beetle** *Ocypus olens*. This is a common garden beetle, well known for its threat display of erecting the hind part of its body and opening its mandibles wide. It may be found variously among vegetable refuse, under stones, or merely wandering about in the open. 2 cm.

Stag Beetle *Lucanus cervus* (1). A well-known and striking insect, the largest British beetle, named from the similarity of the male's mandibles to a stag's antlers. It is mainly southern in distribution and is especially common in the London suburbs. Male without horns 3.5 cm, with horns 5 cm; female 3 cm. The **Lesser Stag Beetle** *Dorcus parallelipipedus* has very much smaller 'antlers', resembling those of the female Stag Beetle.

Dor Beetle *Geotrupes stercorarius*. A common dung beetle, noted for its bumbling habit of colliding with walkers on warm summer evenings. It is also known popularly as Shardborne Beetle, Dumbledor, the Clock, and the Lousy Watchman. The last of these names arises from its often being covered with tiny mites, *Gamasus coleopterorum*. 2 cm.

Cockchafer or **Maybug** *Melolontha melolontha*. One of our best-known beetles, familiar for flying around on late spring evenings making a loud humming sound. Its larvae can be destructive agricultural pests, and are known to farmers as White Grubs or Rookworms. 2.5 cm.

Rose Chafer or **Rose Beetle** *Cetonia aurata*. A handsome golden green beetle, allied to the dung beetles, it is sometimes seen in roses and other large garden flowers. It is widespread but local throughout Britain and Ireland. 1 cm.

Garden Chafer *Phyllopertha horticola*. A common beetle, often seen in large swarms on a sunny day, that can be a destructive garden pest, especially damaging to the leaves of fruit trees. It is also well known to anglers, who sometimes call it by the colloquial names of Bracken-clock or Coch-a-bonddu, since it is a favourite food of the Brown Trout. 8 mm.

237

Soldier Beetle *Rhagonycha fulva*. A common small beetle which is often to be found on wild flowers, especially Hogweed and other umbellifers. Its name derives from its colour, and refers to the fact that until this century soldiers in the British Army wore red uniforms. 8 mm.

Glow Worm *Lampyris noctiluca*. A beetle which is noted for the light-producing organs with which the larva-like females attract the males at night. These organs produce light by oxidising a compound called luciferin and reflecting it by means of minute urate crystals, and are therefore luminous, not phosphorescent. The Glow Worm is mainly southern in distribution. Male 1–1.2 cm, female 1.2–1.8 cm.

Larder Beetle, Lard Beetle or **Bacon Beetle** *Dermestes lardarius*. A scavenging beetle which can be a serious pest of bacon, hides and fatty foodstuffs. It can do further damage at the larval stage when the grub is seeking a place to pupate, because it burrows into all kinds of vegetable substances nearby, such as books, cork or woollen clothing. 8 mm.

Museum Beetle *Anthrenus museorum*. A relative of the Larder Beetle, and likewise a scavenger of dead animal material, it is infamous for the damage it can do to museum collections, especially of butterflies, moths and other insects. The adult beetle may be found on flowers, especially on umbellifers. 4–5 mm.

Carpet Beetle *Attagenus pellio*. Another scavenging beetle which is an indoor pest because its larvae attack skins, furs and textiles such as horse blankets. It can also destroy entomological specimens, like the Museum Beetle (above). It is a very common pest in both houses and warehouses, and an unwelcome invader of corn mills. 5–6 mm.

Furniture Beetle or **Woodworm** *Anobium punctatum*. A common wood-boring beetle, notorious for its destructiveness to wooden beams in old houses, and to furniture. In fact it accounts for some 80 per cent of all insect damage to timber in buildings in Britain. It also occurs out of doors throughout Britain and Ireland; its larvae may be found in old tree-stumps and other dead wood. 4–5 mm.

larvae damage

Death Watch Beetle *Xestobium rufovillosum*. A furniture beetle, well known for the tapping sound made by the adults before they emerge from the pupal chamber. In former times it was liable to be heard doing this in old wooden houses by people watching at the bedsides of sick people – hence its name. The burrowing larvae can inflict immense damage to timber construction, as they once did, for instance, to Westminster Hall in London. 6–7 mm.

Cardinal Beetle *Pyrochroa coccinea*. One of 3 bright red beetles so named because their colour resembled that of the red robes of a cardinal. The larvae are wood-feeders and the adults may be found under bark, but also visit flowers. 1.5–1.7 cm.

Oil Beetle *Meloë proscarabeus*. One of a group of 7 wingless beetles, named from their habit of discharging their evil-smelling, oily yellow blood as a defence mechanism. Their larvae, called triangular larvae, feed on the food stored in the nests of certain bees and look so like lice that they were originally classified as such by entomologists. 2–3 cm.

Mealworm Beetle *Tenebrio molitor*. This is the beetle whose larvae, known as yellow mealworms, are a great pest of corn, flour and other stored products. However they can also be useful to man and are reared in large numbers by naturalists, zoo keepers and others for feeding to insectivorous mammals and birds. 1.5 cm.

larva

Churchyard Beetle or **Cellar Beetle** *Blaps mucronata*. One of the nocturnal ground beetles, quite widespread in Great Britain, this species is often found at night in cellars, stables and large kitchens. Its larvae feed on decaying vegetable matter, in such obscure places as the cracks between floorboards. 2 cm.

Elm-bark Beetle *Scolytus scolytus*. This is the insect that has helped to destroy the greater part of the Elm population of England in recent years, thus changing the face of the countryside irreparably. It inhabits the bark of the trees, where its branching galleries can usually be seen, and in so doing spreads the agent of Dutch Elm Disease, the fungus *Ceratostomella ulmi*. This destructive beetle will also inhabit other trees, such as Oak, Ash and Poplar. 4–5 mm.

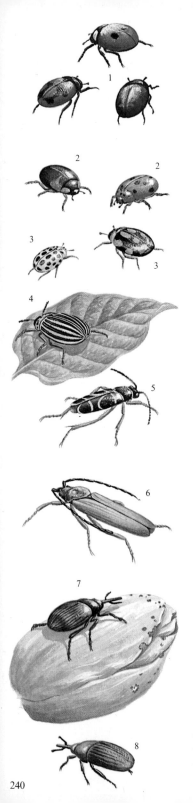

Two-spot Ladybird *Adalia bipunctata* (1). Ladybirds are a family (Coccinellidae) of small, usually red or yellow, beetles with black spots. They are well-known friends of the gardener because both adults and larvae feed on aphids and scale insects. Care should therefore be taken not to destroy them when spraying plants with insecticide. This species is very variable, as shown by the illustrations. 5–7 mm. **Ten-spot Ladybird** *Adalia decempunctata* (2) and **Twenty-two-spot Ladybird** *A.22-punctata* (3) are 2 more fairly common species of Ladybird. 5–7 mm and 3–5 mm respectively.

The alien **Colorado Beetle** *Leptinotarsa decemlineata* (4) from N America, striped not spotted, would become a serious pest of potato crops if it were allowed to establish itself in Britain. If seen, it should be captured and reported promptly to the local police station. 1–1.2 cm.

Wasp Beetle *Clytus arietis* (5). A common beetle named from its resemblance to a Common Wasp, which it exploits, when flying near flowers in the sunshine or running over tree-trunks, by imitating its movements, such as a nervous tapping of the antennae. 1.5–1.8 cm.

Musk Beetle *Aromia moschata* (6). A handsome bright green (occasionally bright blue or even coppery) longhorn beetle, both sexes of which emit a strong, musky odour. This species is widespread throughout Britain and Ireland in the neighbourhood of willows and sallows, emerging in July and August. 3 cm.

Appleblossom Weevil *Anthonomus pomorum* (7). One of a very large family of small beetles, in fact the largest family in the Animal Kingdom – the Cucurlionidae; there are more than 500 species in Britain alone. This species can be a tiresome and destructive pest of apple and pear orchards, the larvae feeding on the unopened buds of the blossom. It has its own specific ichneumon predator.

Grain Weevil *Sitophilus granarius* (8). This, together with the **Rice Weevil** *S. oryzae*. is a serious pest of stored grain. It can be either sooty black or tinged with a rusty colour. The Rice Weevil is distinguished by 2 pale patches on each wing-case. Grain Weevil 2–4 mm, Rice Weevil 2–3 mm.

Pond Sponge *Spongilla lacustris* (1) and **River Sponge** *S. fluviatilis* (2). Our 2 common freshwater sponges are somewhat misnamed, for the Pond Sponge is usually found in the deeper parts of slow-moving rivers, while the River Sponge occurs in lakes and ponds. The larvae of a certain species of 2-winged fly, the Sponge Fly *Sisyra fuscata*, live on or in freshwater sponges, sucking their juices.

Jenkins' Spire Shell *Potamopyrgus jenkinsi*. This species is a small, dark, freshwater snail that was until about 90–100 years ago confined to brackish water in estuaries and around the coast. Since then it has inexplicably spread to fresh water all over Britain, and is now common in rivers and streams. H 4–6 mm.

Great Pond Snail *Limnaea stagnalis*. This large freshwater snail can be up to 5 cm long and eats both animal and vegetable matter; it has even been known to attack newts and small fish. The Great Pond Snail is widespread and common throughout Britain in larger ponds, and sometimes occurs also in canals and slow-flowing rivers. H 8–12 mm.

River Limpet *Ancylastrum fluviatilis* (3). A common freshwater mollusc in streams and small lakes in most parts of Britain, it attaches itself to stones or stout vegetation. 6–9 mm. The **Lake Limpet** *A. lacustris* is less common, with a flatter shell, and usually prefers still water. 6–7 mm.

Ramshorn Snail or **Trumpet Snail** *Planorbis planorbis* (4). Often kept in aquaria, these freshwater snails, about 1.8 cm across, are widespread and common in ponds, ditches and slow-moving water throughout Britain. The larger **Great Ramshorn Snail** *P. corneus* may be as much as 2.5 cm across and is a more local species.

Grove Snail *Cepaea nemoralis*. A common and very variable land snail of woods, hedges, chalk downs and many other habitats. It is widespread and locally common in England, Wales and Ireland, but absent from N Scotland. It is used to demonstrate the survival of the fittest in evolution, because Thrushes eat the snails which are least well camouflaged against their environment. 1.8 cm across.

white-lipped variety

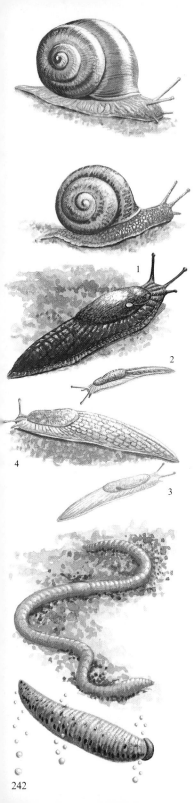

Roman Snail *Helix pomatia*. Considerably the largest British land snail, at 4.5 cm in both length and breadth, this is a local creature of calcareous soils in the S half of Britain. Though the Romans may have relished it, it was in Britain before they invaded the country. This is the snail eaten in France as the *escargot*. It hibernates in winter.

Common Snail *Helix aspersa*. This widespread and common garden snail is very variable in appearance. In Scotland it does not occur N of the Moray Firth. It appears to be greatly affected by atmospheric pollution, and is scarce in or near smoky industrial towns. Like the Roman Snail, it hibernates during the winter months. It is, or was, eaten in several parts of England. H 3.5 cm, and 3.5 cm in breadth.

Black Slug *Arion ater* (1). Slugs are closely related to snails, but have a soft shell concealed under their mantle. This is the large black slug so unwelcome to gardeners when it appears on the lawn after a heavy shower. It may also have a yellowish rufous form, and is rather variable. Also found in woods and fields. Both the Black Slug and the **Garden Slug** *A. hortensis* (2) are widespread and common.

Field Slug *Agriolimax agrestis* (3). A widespread and common slug, found in all kinds of habitats, including gardens and farmland. It can be a great pest to both farmers and gardeners. 3·5 cm. The **Yellow Slug** *Limax flavus* (4) is widespread and fairly common in England, Wales and Ireland, but less so in Scotland, often appearing in and around houses, feeding on decaying vegetation, not green shoots. 7.5–10 cm.

Common Earthworm *Lumbricus terrestris*. Earthworms may number millions per acre of soil, performing a vital function by aerating the soil and moving nutrient salts to the surface. Attempts to kill earthworms to prevent their casts appearing can only harm the fertility of a lawn. *L. terrestris* is one of our commonest species, and can grow up to 30 cm in length.

Horse Leech *Haemopis sanguisorba*. Leeches are worm-like aquatic animals, with a sucker at each end so that they can both grip their victims and anchor themselves in swift currents. They are mostly bloodsuckers, and secrete an anti-coagulant, so that a leech wound takes a long time to heal. Despite its name, the Horse Leech feeds on frogs and fishes. Up to 6 cm.

Pill Millipede *Glomeris marginata*. A common millipede, especially on calcareous soils, where it may be found in fields and hedges and among fallen leaves. It looks remarkably like the Pill Woodlouse (below), but in fact has certain differences such as a shinier cuticle, shorter antennae, and 10 more pairs of legs. Up to 2 cm.

Common Centipede *Lithobius forficatus*. This, the most frequently seen centipede, is widespread and common in England, but less so in Scotland, Ireland and Wales. It can often be seen in and near houses and gardens, both in the country and in the suburbs, and in addition frequents woods, grassland and moorland. A typical place to find a Common Centipede is under a stone. Up to 3 cm.

Water Slater, Hog Slater or **Water Louse** *Asellus aquaticus*. A common small scavenging freshwater crustacean of streams and ponds, especially water that is recovering from having been polluted by sewage. *A. meridianus* is another common species. Both are related to the woodlice, and are widespread. Up to 2 cm.

Pill Woodlouse or **Pill Bug** *Armadillidium vulgare*. Woodlice are among the few terrestrial crustaceans, and live in mainly damp places, such as under stones or logs. This species is particularly frequent on calcareous soils. It has a habit of rolling into a ball, and owes its name to having been used as a pill by quack doctors in earlier times. Up to 1.8 cm.

Common Woodlouse *Oniscus asellus*. This and *Porcellio scaber* are our 2 commonest woodlice. Both are frequent, often in clusters, under stones, bark and logs. *Porcellio* may also go quite high up in a tree. Neither rolls up in a ball like the Pill Woodlouse. Up to 1.7 cm.

Freshwater Shrimp *Gammarus pulex*. A very common small freshwater crustacean of rivers and streams throughout Britain and Ireland. It is not at all closely related to the marine shrimps, being an amphipod, not a decapod. The Freshwater Shrimp feeds as a scavenger, but will also prey on smaller animals. It progresses rapidly but jerkily by swimming. Up to 3 cm.

Fairy Shrimp *Cheirocephalus diaphanus*. A transparent freshwater crustacean, about 2.5 cm in length, that swims on its back and is found in the kind of temporary pool, including even muddy cart tracks, that dries up in summer. It is not closely related to either the Freshwater Shrimp or the Common Coastal Shrimp.

Crayfish *Astacus pallipes*. Our largest freshwater crustacean, not unlike a miniature Lobster, to which it is closely related. The Crayfish is widespread but decreasing in rivers and streams that flow fast enough to be well oxygenated, and prefers calcareous water, meaning water that flows from or over chalk and limestone rocks. It is still prized as a delicacy in some country districts. Up to 10 cm.

Trapdoor Spider or **Purse-web Spider** *Atypus affinis*. Our only spider related to the tropical trapdoor spiders, unlike which it does not make a hinged door to the closed silken tube within which it lives. It seizes its prey with its exceptionally long jaws while the prey is resting on the aerial portion of the tube, hauling it inside through a rapidly made slit. 1.2 cm.

Water Spider *Argyroneta aquatica*. Our only spider that lives in the water, constructing an underwater bell which it fills with air bubbles trapped in its body hairs. It hunts various kinds of small aquatic animal, and may eat them either under water or on land. Frequent in ponds and ditches in the S half of England. 1 cm.

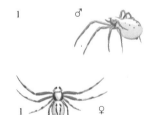

Crab Spider *Misumena valia* (1). One of a group of spiders named from the crab-like stance and sideways movements they adopt, the result of their having curved legs. This species is commonly found in bushes in the S of England, and its egg cocoon may be found in a folded leaf. 8–12 mm.

Zebra Spider *Salticus scenicus*. This species is a small spider that is named from its black and white stripes. It is widespread and common over most parts of Britain and is most often seen on walls and fences, though not infrequently it also ventures indoors. 6–7 mm.

Daddy-long-legs Spider *Pholcus phalang-oides*. A long-legged spider named from its resemblance to the Crane-flies, which are also known as Daddy-long-legs. It inhabits houses in the S of England and S Wales, and indeed is almost universally present along the coast. The Daddy-long-legs Spider often suspends itself between the walls and the ceiling. 4–5 mm.

Mesh Web Spider *Ciniflo fenestralis*. One of a small group of spiders whose tangled patches of rough silk web are familiar in outbuildings and cellars, often being woven across the window panes, and also on walls, posts and palings. This species is rather commoner in the N and more often seen out of doors than the other 2, which are the widespread *C. similis* and the large *C. ferox*. 1.2–1.5 cm.

House Spider *Tegenaria domestica* (1). The commonest of 3 large, long-legged spiders found indoors, especially in sheds and cellars. The 2 others, *T. atrica* and *T. parietina*, are both larger. *T. parietina* is sometimes called Cardinal Spider, from the legend that it once frightened Cardinal Wolsey at Hampton Court. It is about 1.75 cm long, while its longest legs may measure more than 5 cm.

Steatodea bipunctata. This species is another small spider that is very common throughout Great Britain in crevices and corners in outhouses, such as stables and garden sheds. Its web is much less complex than those of the orb-web spiders, consisting of no more than a tangle of criss-crossing threads. 1.2 cm.

Wolf Spider *Lycosa amentata* (2). One of a group of spiders named from their habit of running down their prey on foot instead of waiting for it to enter a web. This species is common on forest floors. The commonest species is *L. saccata* (3), which is often seen running on the ground in both woods and farmland. 8–10 mm.

Pirate Spider *Pirata piratica*. Another species of wolf spider that is especially common by lakes and ponds, pursuing its prey not only on the banks but also over the surface film of the water. Its method of detecting the presence of prey in its vicinity is indicated by the fact that a tuning fork vibrated experimentally will bring it out in a rush. 6–8 mm.

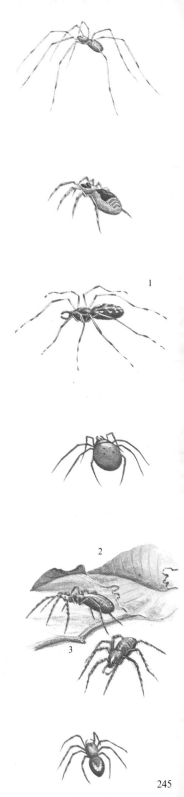

Micrommata viridissima. The male, illustrated here, is one of our most brightly coloured spiders, but the female is all green with merely a spear-shaped darker mark on her abdomen. This is a woodland spider, scattered widely over England and Wales, that does not spin webs but uses its green camouflage to waylay unsuspecting insects. 1.1–1.3 cm.

♂

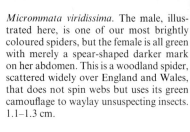

Cyclosa conica is a small – up to 6 mm – spider of such a remarkable shape that its protruding abdomen appears to give it a tail. This species spins a small-meshed web, with a band across the centre in the middle of which, partially camouflaged, the spider rests, awaiting its prey.

Money Spiders Linyphiidae. A large family of tiny spiders, extremely common – over a million have been estimated in one Sussex field – with hammock webs that become very conspicuous on lawns after autumn dews. *Linyphia triangularis*, illustrated here, is abundant, especially in pinewoods, where its webs may be scattered over large areas. 6–8 mm.

Aranea cucurbitina. One of our beautiful small spiders, this species grows up to 6 mm in length and has a green abdomen and reddish brown cephalothorax ('head'), but these lovely colours soon fade after it dies. It is very common in woods, hedgerows and bushes, and in gardens is reputed to be especially attracted to rose trees. Like the following species, it spins the advanced type of web known as an orb-web.

Garden Spider *Aranea diadema.* One of our commonest spiders, this species, despite its name, is no means confined to gardens. The large white cross on its brown back is diagnostic. It spins orb-webs, which are the most sophisticated form of spider's web, by laying spirals of silk around a series of radial threads, which are themselves attached to an outer framework of threads. 1.2–1.5 cm.

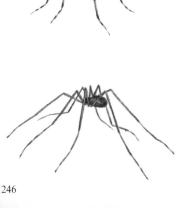

Harvestman or **Harvest Spider** Opiliones. A group of very long-legged, predatory relatives of the spiders, with an undivided body and no silk glands – therefore they are not able to spin webs. *Phalangium opilio*, the species illustrated here, is one of the commonest of them, and is especially conspicuous in the autumn. 3–6 mm.

The Seashore

The sea is a totally different biological environment from both land and fresh water – the necessity to adapt to the highly saline content of sea water means that very few species are common to both. Fish that ascend rivers to spawn, such as Salmon, Sea Trout and the lampreys, are among the few that spend part of their life in both fresh and salt water. A few other animals, such as the Three-spined Stickleback, may be found in either fresh and brackish or in brackish and salt water.

The seashore has a very different array of species from the open and deep sea. The shore is dominated by the tides, and its animals and plants have to be adapted to the varying degrees of exposure to salt water and fresh air that the twice-daily regime of the tides imposes. At the very top of the shore is the **splash zone**, which may get drenched with spray during a storm with an onshore wind every now and then. Below that is the **upper shore**, which is covered only by high tides above the average high-tide level; then the **middle shore**, between average high-tide and average low-tide levels; and finally the **lower shore**, below the average low-tide level, where only a few hours a year may have to be spent exposed to the air. The **sub-littoral fringe**, the area below the extreme low-water level of the spring tides, is never exposed at all. Spring tides are the highest tides, at full and new moon, when sun and moon are pulling together, once a fortnight; at the opposite extreme are the neap tides, also fortnightly.

The main types of shore are **rocky**, with many rock pools and weed-covered rocks, that form the most rewarding hunting ground for the marine naturalist; **shingly**, with millions of small, rounded pieces of rock called pebbles; and **sandy**, where the rock particles have been ground much finer by the perennial action of the sea. Each of these types has its own characteristic flora and fauna, though there is little flora except on the rocks.

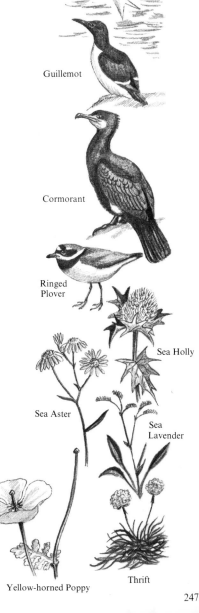

Black-headed Gull

Common Tern

Guillemot

Cormorant

Ringed Plover

Sea Holly

Sea Aster

Sea Lavender

Sea Sandwort

Yellow-horned Poppy

Thrift

Sea Lettuce or **Green Laver** *Ulva lactuca*. A common green seaweed with broad, supposedly lettuce-like leaves, left sprawling when the tide goes out on rocky shores. It is especially common in the summer and in places where fresh water flows out, especially if this water happens to be slightly polluted with sewage.

Oarweed *Laminaria digitata*. One of the commonest of the large brown seaweeds that are collectively called tangle in Scotland, and when thrown ashore are called kelp and either used as manure or burned to produce a chemically rich ash. This is a very common species on rock-covered lower shores.

Sea Belt *Laminaria saccharina*. This is another species of tangle or Oarweed, easily recognized by the very crinkly edges of its fronds. It can be found attached to small stones and rocks on pools and on mudflats and sandflats from the middle shore downwards.

Bladder Wrack *Fucus vesiculosus* (1). One of the 2 brown seaweeds of the shore that have conspicuous air bladders. In this species the bladders are oval; in the other, **Egg Wrack** or **Knotted Wrack** *Ascophyllum nodosum*, they are round. Both grow on rocks or breakwaters on the middle shore, sometimes well up estuaries in almost brackish conditions.

Serrated Wrack or **Saw Wrack** *Fucus serratus* (2). Another very common brown seaweed, growing in similar places to Bladder Wrack and Egg Wrack. Another similar wrack is **Flat Wrack** *F. spiralis*, which has no serrations on its fronds, and has terminal granular fruiting bodies that look superficially like bladders. It grows lower down the shore, and is usually 3 m deep at high neap tides.

Thongweed *Himanthalis elongata* (3). A brown seaweed with strap-like fronds anything from 3 m to 8 m long, locally common on rocky shores, especially exposed ones. **Bootlace Weed** *Chorda filum* (4) has even narrower fronds, exactly like an old-fashioned leather bootlace. It is very common on the lower shore and below, and of course washed up in the kelp (see Oarweed, above).

Edible Dulse *Rhodymenia palmata*. A widespread and common red seaweed, at one time the staple diet of people living on the W coasts of Scotland and Ireland. It grows on rocky shores, on both rocks and other seaweeds, such as Oarweed. Another dulse that can be eaten is *Dilsea carnosa*, found on small stones on sand-flats.

Breadcrumb Sponge *Halichondria panicea*. Sponges are primitive, plant-like animals, which feed by passing through their gut cavity a stream of water drawn in through their many pores. They have tiny, free-swimming larvae, and also reproduce asexually by means of overwintering buds. This species is common on the lower shore and may also be yellow or orange.

Jellyfish *Aurelia aurita*. Jellyfish spend the greater part of their lives as free-swimming medusae, among the plankton (tiny animals and plants that float on the surface of the sea). This is the commonest species of jellyfish in British seas, often cast ashore after a storm. The largest jellyfish in the world, *Chrysaora isosceles*, also occurs in British seas.

Beadlet Anemone *Actinia equina*. Sea anemones are carnivorous marine animals, like truncated cones of jelly on the rocks, named from the flower-like appearance of their expanded tentacles. This is one of our commonest and most widespread species, varying in colour from red to green, brownish green, yellow–brown and strawberry red with green spots.

Dahlia Anemone *Tealia ferina*. A common, widespread and variable sea anemone, found in rock pools and cracks on the middle and lower shore. It may be green, grey or red, of various shades, and the grey warts are sticky, so that small fragments of stone or shell often adhere to them.

Ragworm *Nereis diversicolor* (1). Marine polychaete or bristle worms are often brightly coloured and some species may grow up to 3 m long. The Ragworm is a common species of brackish water, for instance estuaries. The equally common **Lugworm** *Arenicola marina* (2), whose casts cover the sand-flats at low tide, is much dug by fishermen for bait, being especially good for catching Bass.

249

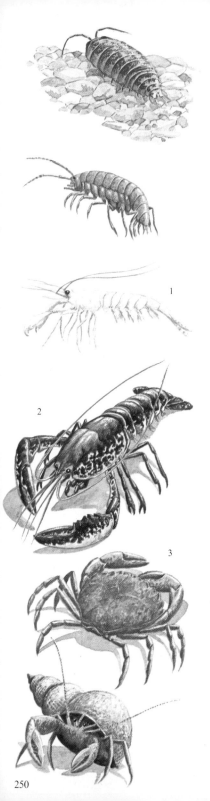

Sea Slater *Ligia oceanica*. An abundant isopod crustacean, related to the terrestrial Woodlice, which may also be found accompanying it in its favourite habitat under stones and seaweed above the high water mark, also in cracks in harbour walls and similar places. The Sea Slater is a fast-running scavenger, active at night.

Sandhopper *Talitrus saltator*. This species is a very common amphipod crustacean, which can be found among rotting vegetation, mostly seaweed, on sandy shores. As its name suggests, it moves by jumping or hopping.

Shrimp *Crangon vulgaris* (1). A common decapod crustacean of shallow pools on sandy shores and estuaries, moving to deeper water in autumn. It is valued as a delicious food for humans. Its larger relative the **Common Prawn** *Leander serratus* is found only on Britain's S and W coasts, where it becomes especially common in late summer or flood tides in pools full of seaweed.

Lobster *Homarus vulgaris* (2). One of our larger decapod crustaceans, the Lobster is more familiar as the red skeleton it becomes after boiling. It frequents deeper waters on rocky shores and is not often seen above the low water mark. The 2 larger claws are usually unequal in size. The **Crawfish** or **Spiny Lobster** *Palinurus vulgaris* has no large claws.

Shore Crab *Carcinus maenas* (3). This is the commonest crab found on the middle and lower shore and in estuaries. It is usually dark green, but sometimes also has a pattern consisting of paler marks. The crab most frequently eaten by humans is the **Edible Crab** *Cancer pagurus*, which is to be found for the most part on rocky seashores.

Hermit Crab *Eupagurus bernhardus*. This common crab has the curious habit of occupying the shells of winkles, whelks and other marine molluscs, its own body being unprotected by a carapace. It must never be pulled out, because it would break in half, but it can be persuaded to release its grip by holding a lighted match under the edge of the shell. The Hermit Crab should then be allowed to return.

Chiton or **Coat of Mail Shell**
Lepidochiton cinereus.
Common between tide-marks,
feeding on seaweeds.

Acorn Barnacle *Balanus balanoides*. The
barnacle that scratches bathers' shins. This
species is commoner in the N, while the
very similar *Chthalamus stellatus* is more
frequent in the S.

White Tortoiseshell Limpet
Acmaea virginea.
Widespread and common,
especially on Oarweed.

imm.
1

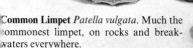

Blue-veined Limpet or
Blue-eyed Limpet (1)
Patella lucida.
Widespread and
common, often on
Oarweed.

ad.
1

Common Limpet *Patella vulgata*. Much the
commonest limpet, on rocks and breakwaters everywhere.

Common Top Shell or **Painted Top Shell**
Calliostoma zizyphinum. One of the most
widespread and common top shells, often
cast ashore.

Grey Top Shell *Gibbula cineraria*.
Widespread and common on
middle and lower shores.

Common Necklace Shell *Natica alderi*. Lives
in the sand and preys on other molluscs.

Rough Periwinkle *Littorina
saxatilis*. Widespread and
common on the middle shore.

Flat Periwinkle *Littorina littoralis*.
Abundant on the lower shore, especially on brown seaweeds (wracks).

Common or **Edible Winkle** or **Periwinkle** *Littorina littorea*. Abundant
on rocks and seaweed from the
middle shore downwards.

251

Banded Chink Shell *Lacuna vincta*.
Widespread on seaweed.

Wendletrap *Clathrus clathrus*. Widespread
and common offshore.

Needle Shell *Bittium reticulatum*. Wi
spread and common on the shore.

Tusk Shell or **Elephant Tooth Shell** *Dentalium entalis*. More frequent in the N.

Slipper Limpet *Crepidula fornicata*. A
serious alien pest of oyster beds.

Cowrie *Trivia monacha*. The commonest
European cowrie – most species are tropical.

Pelican's Foot Shell
Aporrhais pes-pelecani.
Common in muddy
gravels offshore.

Tower Shell *Turritella communis*. Also common in muddy gravels offshore.

Thick-lipped Dog Whelk
Nassarius incrassatus.
Common on rocky shores

Netted Dog Winkle *Nassarius
reticulatus*. Widespread and
common on rocky shores.

Common Dog Whelk or **Dog Winkle** *N
lapillus*. A barnacle feeder.

Common Whelk or **Buckie** *Buccinium
datum*. See Strandline, p. 256.

Common Mussel *Mytilus edulis*. Widespread and gregarious on the shore.

Great Scallop, Edible Scallop or **Clam** *Pecten maximus*. A somewhat neglected delicacy.

Common Oyster *Ostrea edulis*. A still relished delicacy, mainly found in SE estuaries.

Common Cockle *Cardium edule*. Still eaten in S Wales and elsewhere.

Dog Cockle *Glycymeris glycymeris*. Widespread and common on the shore.

Thin Tellin *Tellina tenuis*. Common on sandy shores.

Banded Carpet Shell or **Pullet Carpet Shell** *Venerupa pullastra*. Common between tide marks.

Sand Gaper *Mya arenaria*. In the USA, eaten as the Soft-shelled Clam.

Common Piddock *Pholas dactylus*. Bores into chalk and other soft rocks.

Razor Shell *Ensis ensis*. Widespread and common.

Common Octopus *Octopus vulgaris* (1). Octopuses hide in rock crevices, emerging to seize their prey with their suckers. *O. vulgaris* is a marine cephalopod mollusc that is quite common on the S coast of England in some years. It has 4 pairs of long tentacles and 2 rows of suckers, differentiating it from the mainly northern **Curled Octopus** *Elodone cirrhosa*, which has only a single row of suckers.

Common Starfish *Asterias rubens*. Starfish are free-living, flattened, star-shaped, carnivorous marine animals. *A. rubens* is a widespread and common species on the lower shore, and can be quite variable in colour – some examples, for instance, are violet in hue.

Brittle-star *Ophiothrix fragilis*. One of the commonest of a group of starfish-like marine animals, characterised by a fairly clear division between the central disc or body and the arms, of which there are usually 5. It is found on the lower shore, and is extremely brittle, its arms often fragmenting into pieces if touched.

Common Sea Urchin *Echinus esculentus*. Sea Urchins are spiny marine animals, roundish, oval or heart-shaped, with a brittle, limy external skeleton called a test. This is one of the commonest and most widespread species, inhabiting the lower shore. As its Latin name suggests, the Common Sea Urchin is edible.

Lesser Sand-eel *Ammodytes lancea* (2). This and the **Greater Sand-eel** *A. lanceolatus* are 2 small, eel-shaped fish of sandy shores, not related to the true eels. Both of these species are widespread and common, and are much preyed on by terns, which can often be seen carrying them off to feed to their young.

Common Goby *Gobius minutus*. Gobies are small fishes of rocky and sandy shores. *G. minutus* is one of the commonest of the 13 species found on British and Irish shores, often extending into the brackish water of estuaries. It is some 5–7 cm in length, and has its pelvic fins (the fins on its belly) modified to form a sucker.

Greater Weever or **Sting-fish** *Trachinus draco* (1). This species and the **Lesser Weever** *T. vipera* (2), which is much the commoner, are small marine fish that lie half-concealed in the sand in shallow water, and can inflict a painful wound with their poisonous spines on any unsuspecting bather who accidentally treads on them.

Common Blenny or **Shanny** *Blennius pholis* (3). There are 5 species of blenny, all superficially goby-like, in British inshore waters. This species is much the commonest, and can be found in various shades of olive, green and yellow. The much less common **Butterfly Blenny** *B. ocellaris*, of S and W coasts, is named from the large 'eye' on its dorsal fin, like that of a Peacock Butterfly.

WHALES, PORPOISES AND DOLPHINS Cetacea

Common Porpoise *Phocaena phocaena*. The smallest and commonest cetacean in British and Irish seas, often seen offshore in small schools, characteristically surfacing 2 or 3 times and then disappearing. It can be distinguished from the dolphins by its blunt head and small, triangular back fin. It is not usually above 2 m long. A slow swimmer, it does not often leap from the water.

Bottle-nosed Dolphin *Tursiops truncatus* (4). This and the smaller **Common Dolphin** *Delphinus delphis* are the 2 commonest dolphins off our coasts, differentiated from porpoises by their pointed beaks, sharply pointed back fins and habit of leaping from the water. The Bottle-nosed Dolphin is stouter, less slender and less brightly coloured, and usually goes in smaller schools.

Killer Whale *Orcinus orca*. The largest of the dolphins, the males growing up to 9 m long, are among the more easily recognisable cetaceans, with their striking black and white body pattern and their tall, sharp back fin. Among the most ferocious animals in the sea, they prey on smaller whales, dolphins, porpoises, seals, Salmon and other large fish.

Basking Shark or **Sailfish** *Cetorhinus maximus*. This completely harmless animal – it feeds on plankton – is not only the largest European fish, but the second largest fish in the world. It basks in the sunshine on the surface of the water, and is quite common on W coasts.

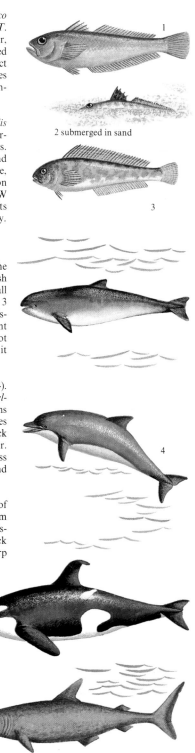

2 submerged in sand

3

4

THE STRANDLINE

Many fascinating objects get washed up by the sea, and stay at the furthest point of the highest tides, known as the strandline. An example is a hard, brown bean, not unlike a broad bean, which is the fruit of the tropical climber *Entada gigas*. These beans will have been brought across the Atlantic from the West Indies by the Gulf Stream.

Gribble *Limnoria lignorum* is an isopod crustacean, resembling the woodlice so commonly found on land, that burrows into wood. Pieces of wood found on the strandline often contain large numbers of holes 5–6 mm across. These have been bored by the Gribble, females of which may sometimes be found at the end of the burrows.

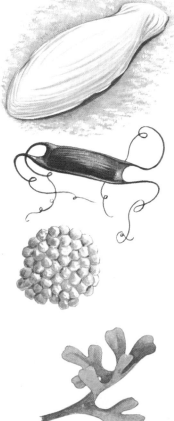

Cuttlefish Bones, illustrated here, are the internal shell of the **Common Cuttlefish** *Sepia officinalis*, a cephalopod mollusc related to the octopuses and squids, which comes close inshore to spawn in summer, especially along the S coast of England. The dark fluid secreted by the Cuttlefish when threatened is used to make the brown pigment called sepia. Cuttlefish bones are often given to cage birds, to sharpen their beaks.

Mermaid's Purses are the normally empty egg cases of dogfish or rays. That of the **Lesser Dogfish** *Scyliorhinus caniculus* is about 5 cm long, and that of the **Greater Dogfish** *S. stellaris* about 10 cm – in both cases this measurement excludes the 4 tendrils. Mermaid's Purses with points instead of tendrils have come from rays.

Whelk Egg Cases. Clusters of what at first sight appear to be dried masses of frog-spawn are frequently found on beaches at low tide. These are in fact also egg cases, in this instance those of the **Common Whelk** *Buccinium undatum*, whose shell is illustrated on p. 252.

Hornwrack *Flustra foliacea*. What seem to be pieces of dried seaweed on the beach, but are flat and brittle and pitted with minute honeycombs, are the remains of the curious animals known as **Sea-mats** *Bryozoa*. They are actually colonies of tiny animals, each of which would originally have occupied one cell of the honeycomb.

Farm Crops and Livestock

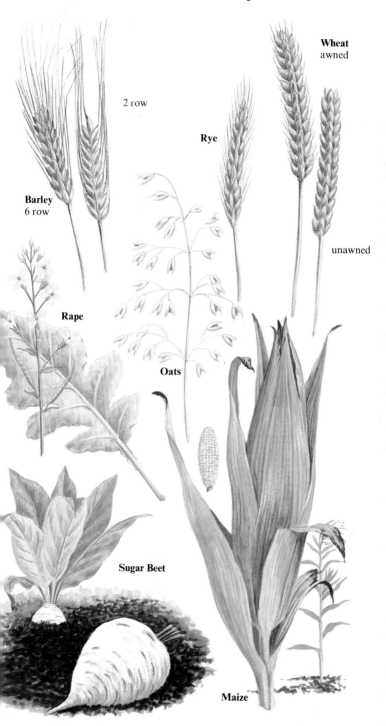

2 row

Wheat
awned

Rye

Barley
6 row

Rape

Oats

unawned

Sugar Beet

Maize

Galloway

Aberdeen Angus

Highland

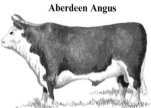

Hereford

Charollais

Devon

Ayrshire

Red Poll

British Friesian

Dairy Shorthorn

Jersey

Guernsey

SHEEP Bovidae

Blackface

Suffolk

Shropshire

Border Leicester

Dorset Horn

Southdown

Cheviot

Romney

HORSES Equidae

Welsh Mountain

Soay

New Forest Pony

Dartmoor Pony

Exmoor Pony

Shetland Pony

Welsh Mountain Pony

259

Glossary

Algae   A group of lower plants including the seaweeds.

Amorphous   Shapeless.

Amphibians   A group of lower vertebrates, including frogs, toads and newts.

Amphipods   A group of small, shrimp-like creatures.

Annual   A plant that lives for a single growing season only.

Antenna   The feeler, of which every insect has two on its head, forming the insect's organs of smell and touch.

Anther   The small sac on the end of the male reproductive structure (the stamen, *q.v.*) in a flower, that contains the spores.

Arthropods   The invertebrate phylum which includes the insects, crustaceans and arachnids (spiders and mites).

Barbel   A fleshy filament hanging from a fish's mouth.

Biennial   A plant that takes two years to complete its life cycle.

Bifid   Split in two.

Bivalve   A mollusc with two shells.

Bract   A small leaf or scale immediately below a flower or cluster of flowers.

Bulbous   Describes a plant growing from an underground bulb; alternatively one that is swollen like a bulb.

Calcareous   Containing calcium in the form of chalk or lime.

Carapace   The upper part of the shell of a tortoise, turtle or crab.

Carnivore   A flesh-eater; *adj.* carnivorous.

Cephalopods   A specialised group of molluscs, including the octopuses and squids.

Character   A distinctive feature.

Chlorophyll   The green colouring matter of almost all plants.

Climbing   Describes a plant that ascends other vegetation, walls etc. by twining or clinging (*cf.* scrambling).

Concentric   Having a common centre, as, *e.g.*, the rings in a section of tree-trunk.

Coniferous   Cone-bearing.

Creeping   A plant whose stems run along the ground, often rooting.

Crustaceans   A group of arthropods, mainly marine, but some living in fresh water and a few on land, including shrimps, crabs and woodlice.

Cultivar   A variety produced in cultivation.

Cuticle   The outermost layer of skin in animals; the hard external case of an arthropod; or the waxy layer on the surface of leaves.

Dabbling duck   A duck that feeds by dabbling on the surface of the water.

Decapods   A group of crustaceans that includes the crabs, lobsters, shrimps and prawns.

Deciduous   With leaves falling each autumn.

Diagnostic   Assisting in diagnosis, or identification.

Diurnal   By day.

Dune-slack   A damp hollow among sand-dunes.

Elytron (*pl.* **elytra**)   The forewing of a beetle.

Evergreen   With leaves that do not fall each autumn.

Feral   Running wild.

Field mark   A character (*q.v.*) that helps identification in the field.

Floret   A small individual flower in a compound head.

Gall   An abnormal growth of plant tissue, usually caused by insects or mites.

Gelatinous   Jelly-like.

Genus   The level of classification between the family and the species.

Gill   The parallel flaps beneath the cap of a toadstool that bear the spores.

Gland   A small sac secreting liquid, often aromatic.

Hawk   To pursue insects as a hawk does its prey.

Herbivore   An animal that feeds on plants; *adj.* herbivorous.

Hermaphrodite   An organism having both male and female reproductive organisms in one body.

Honeydew   A sweet-tasting substance exuded by aphids and other plant bugs.

Host   The animal or plant on which a parasite inflicts itself.

Hybrid   The progeny of animal or plant parents of different genera, species or subspecies.

Insectivore   An insect-eater; *adj.* insectivorous.

Isopods   A group of crustaceans that includes the woodlice.

Keel   The lower part of a peaflower.

Lek   The display ground used for the communal courtship of certain birds and other animals.

Ley   Grass sown as a crop.

Microclimate   The highly localised climate of a strictly limited habitat.

Midrib   The central rib of a leaf, from which the veins spread.

Naked   Describes a plant stem with no leaves.

Omnivore   An animal that eats both animal and vegetable foods; *adj.* omnivorous.

Ovipositor   A specialised egg-laying organ in insects.

Palmate   Describes a leaf with 5 leaflets.

Parasite   An animal or plant living in or on another animal or plant, and feeding on its host (*q.v.*).

Perennial   A plant that continues to grow from year to year.

Petals   The inner row of flat structures surrounding the reproductive centre of a flower, often brightly coloured.

Photosynthesis   The process by which plants synthesise carbohydrates from water and carbon dioxide, using energy derived from sunlight with the aid of chlorophyll (*q.v*).

Pinnate   Describes a compound leaf with several pairs of distinct leaflets.

Pollard   A tree which has been beheaded at about 2 m from the ground.

Predator   An animal that preys on another animal.

Prehensile   Capable of grasping.

Prostrate   Flat on the ground.

Pustulate   Covered with wart-like excrescences.

Rhizome   Underground storage stem.

Runner   A creeping stem above ground, that roots at the tip.

Scavenger   An animal feeding on waste matter.

Scrambling   Describes a plant that ascends other vegetation loosely, without making a strong contact.

Social   Living in a group.

Spadix   A tight spike of tiny flowers.

Spathe   A large bract enveloping one or more flowers.

Speculum   A distinct and differently coloured patch of feathers on a duck's wing.

Stamen   The male reproductive structure in a flower, comprising a thin filament topped by an anther (*q.v.*).

Standard   The upper petal of a peaflower.

Stigma   The receptive tip of the female reproductive structure in a flower, the style (*q.v.*).

Stipule   A small, often leaf-like appendage at the base of a leaf-stalk.

Stock   The base of a plant stem, often woody.

Stolon   *See* runner; *adj.* stoloniferous.

Stridulation   The mechanical sound made by male grasshoppers and crickets.

Style   The female reproductive structure in a flower, comprising a filament and a stigma (*q.v.*).

Succulent   A fleshy plant.

Sucker   A shoot arising from a tree or shrub root.

Superfamily   A group of families that are closely related.

Tail coverts   The feathers at the base of a bird's tail.

Tendril   A twining hair, on the end of a leaf, used by scrambling plants.

Terrestrial   Living on land.

Thallus   The vegetative body of a lower plant; *adj.* thallose.

Trefoil   Describes a leaf with 3 leaflets, as, *e.g.*, clovers.

Umbel   An umbrella-shaped compound flower-head.

Undershrub   A low, woody plant.

Ungulate   A herbivorous hoofed mammal.

Further Reading

Trees

Alastair Fitter and David More, *Gem Guide to Trees*, Collins.

Alan Mitchell, *A Field Guide to the Trees of Britain and Northern Europe*, Collins.

Alan Mitchell and John Wilkinson, *Handguide to the Trees of Britain and Northern Europe*, Collins.

Wild Flowers

Richard Fitter, *Finding Wild Flowers*, Collins.

Richard Fitter and Marjorie Blamey, *Gem Guide to Wild Flowers*, Collins.

Richard Fitter and Marjorie Blamey, *Handguide to the Wild Flowers of Britain and Northern Europe*, Collins.

Richard Fitter, Alastair Fitter and Marjorie Blamey, *The Wild Flowers of Britain and Northern Europe*, Collins.

John Gilmour and Max Walters, *Wild Flowers*, New Naturalist No. 5, Collins.

David McClintock and Richard Fitter, *The Pocket Guide to Wild Flowers* (includes ferns), Collins.

Lower Plants and Fungi

Kenneth L. Alvin, *The Observer's Book of Lichens*, Warne.

F. L. Brightman and B. E. Nicholson, *The Oxford Book of Flowerless Plants*, Oxford University Press.

Morten Lange and F. Bayard Hora, *Collins Guide to Mushrooms and Toadstools*, Collins.

John Ramsbottom, *Mushrooms and Toadstools*, New Naturalist No. 7, Collins.

Mammals

Nicholas Arnold, Gordon Corbet and Denys Ovenden, *Handguide to the Wild Animals of Britain and Europe* (includes reptiles and amphibians), Collins.

F. E. van den Brink, *A Field Guide to the Mammals of Britain and Europe*, Collins.

John A. Burton and Bruce Pearson, *Gem Guide to Wild Animals*, Collins.

M. J. Lawrence and R. W. Brown, *Mammals of Britain: Their Tracks, Trails and Signs*, Batsford.

Birds

Richard Fitter, *Collins Guide to Bird Watching*, Collins.

Richard Fitter and R. A. Richardson, *The Pocket Guide to British Birds*, Collins.

Richard Fitter and R. A. Richardson, *The Pocket Guide to Nests and Eggs*, Collins.

Colin Harrison, *A Field Guide to the Nests, Eggs and Nestlings of British and European Birds*, Collins.

Hermann Heinzel, Richard Fitter and John Parslow, *The Birds of Britain and Europe with North Africa and the Middle East*, Collins.

Hermann Heinzel and Martin Woodcock, *Handguide to the Birds of Britain and Europe*, Collins.

Richard Perry and Martin Woodcock, *Gem Guide to Birds*, Collins.

Roger Peterson, Guy Mountfort and P. A. D. Hollom, *A Field Guide to the Birds of Britain and Europe*, Collins.

Lower Vertebrates

Nicholas Arnold and John A. Burton, *A Field Guide to the Reptiles and Amphibians of Britain and Europe*, Collins.

Bent J. Muus and Preben Dahlstrom, *Collins Guide to the Freshwater Fishes of Britain and Europe*, Collins.

Peter Miller and James Nicholls, *Handguide to the Fishes of Britain and Europe*, Collins.

Butterflies and Moths

Robert Goodden, *British Butterflies: A Field Guide*, David and Charles.

L. G. Higgins and N. D. Riley, *A Field Guide to the Butterflies of Britain and Europe*, Collins.

Richard South, *The Moths of the British Isles* (covers the caterpillars of both butterflies and moths), Warne.

Michael Tweedie and John Wilkinson, *Handguide to the Butterflies and Moths of Britain and Europe*, Collins.

Other Invertebrates

Michael Chinery, *A Field Guide to the Insects of Britain and Northern Europe* (includes butterflies and moths), Collins.

Arnold Darlington, *The Pocket Encyclopaedia of Plant Galls in Colour*, Blandford.

Warne's Wayside and Woodland series has volumes on bees, wasps and ants; beetles; dragonflies; flies; grasshoppers, crickets and cockroaches; land and water bugs; and shells.

The Seashore

John Barrett and C. M. Yonge, *Pocket Guide to the Seashore*, Collins.

Index of English Names

Index of Scientific Names

Fagus sylvatica, 18
Falco columbarius, 161; F. peregrinus,
161; F. subbuteo, 161; F. tinnunculus,
161
Falconidae, 161
Fannia canicularis, 231
Felix sylvestris, 141
Festuca arundinacea, 114; F. gigantea,
116; F. ovina, 114; F. pratensis, 114;
F. rubra, 114
Ficedula hypoleuca, 181
Filipendula ulmaris, 49; F. vulgaris, 49
Fistulina hepatica, 129
Flustra foliacea, 256
Foeniculum vulgare, 65
Fontinalis antipyretica, 124
Forficula auricularia, 223
Formica rufa, 233
Fragaria vesca, 50
Frangula alnus, 25
Fratercula artica, 172
Fraxinus excelsior, 21
Fringilla coelebs, 187; F. montifringilla,
187
Fringillidae, 186–8
Fritillaria meleagris, 100
Fucus serratus, 248; F. spiralis, 248;
F. vesiculosus, 248
Fulica atra, 163
Fulmarus glacialis, 152
Fumaria muralis, 41; F. officinalis, 41
Fumariaceae, 41
Funaria hygrometrica, 123
Fungi, 128–34

Gagea lutea, 100
Galanthus nivalis, 102
Galeopsis speciosa, 77; G. tetrahit, 77
Galinsoga ciliata, 91; G. parviflora, 91
Galium aparine, 73; G. mollugo, 73;
G. odoratum, 73; G. verum, 73
Gallinago gallinago, 169
Gallinula chloropus, 163
Gammarus pulex, 243
Garrulus glandinarius, 189
Gasterophilus intestinalis, 229
Gasterosteidae, 198
Gasterosteus aculeatus, 198
Gavia arctica, 151; G. immer, 151;
G. stellata, 151
Gaviidae, 151
Geastrum triplex, 134
Genista anglica, 52; G. tinctoria, 52
Gentiana nivalis, 71; G. pneumonanthe,
71; G. verna, 71
Gentianaceae, 71
Gentianella amarella, 71; G. campestris,
71
Geoglossum hirsutum, 128; G. viride, 128
Geometra papilionaria, 216
Geraniaceae, 56–7
Geranium columbinum, 57; G. dissectum,
57; G. molle, 57; G. pratense, 57;
G. pyrenaicum, 57; G. robertianum, 57;
G. sanguineum, 57
Gerris lacustris, 225
Geum rivale, 51; G. urbanum, 51
Gibbula cineraria, 251
Glaucium flavum, 41
Glaux maritima, 67

Glechoma hederacea, 76
Glis glis, 137
Glomeris marginata, 243
Glyceria fluitans, 115; G. maxima, 115
Glycymeris glycymeris, 253
Gnaphalium sylvaticum, 90;
G. uliginosum, 90
Gobio gobio, 196
Gobius minutus, 254
Gonepteryx rhaemni, 203
Grifoia gigantea, 129
Grimmia apocarpa, 125; G. pulvinata, 125
Grossulariaceae, 25
Gryllus campestris, 222
Guttiferae, 60
Gymnadenia conopsea, 104
Gyrinus natator, 236

Haematopoda crassicornis, 229;
H. pluvialis, 229
Haematopodidae, 164
Haematopus ostralegus, 164
Haemopsis sanguisorba, 242
Halichoerus grypus, 141
Halimione portulacoides, 34
Hamearis lucina, 203
Hedera helix, 28
Helianthemum nummularium, 61
Helix aspera, 242; H. pomatia, 242
Helleborus foetidus, 38; H. viridis, 38
Hepialus humuli, 209; H. lupulina,
209
Heracleum mantegazzianum, 64;
H. sphondylium, 64
Hesperia comma, 207
Hesperis matronalis, 46
Hieracium aurantiacum, 97; H. murorum,
97; H. pilosella, 97
Himanthalis elongata, 248
Himantoglossum hircinum, 106
Hipparchia semele, 202
Hippocrepis comosa, 54
Hippophae rhamnoides, 26
Hippuris vulgaris, 107
Hirundinidae, 176
Hirundo rustica, 176
Holcus lanatus, 118; H. mollis, 118
Homarus vulgaris, 250
Honkenya peploides, 35
Hordeum europaeus, 117; H. marinum,
117; H. murinum, 117; H. secalinum,
117
Hottonia palustris, 67
Humulus lupulus, 28
Hyacinthoides hispanicus, 101; H. non-
scriptus, 101
Hydrobates pelagicus, 152
Hydrocharis morsus-ranae, 99
Hydrocotyle vulgaris, 63
Hydrometra stagnorum, 225
Hydrophilus piceus, 236
Hylocomium splendens, 124
Hyloicus pinastri, 215
Hymenoptera, 231–5
Hyoscyamus niger, 80
Hypericum androsaemum, 60; H. elodes,
60; H. humifusum, 60; H. perforatum,
60; H. pulchrum, 60
Hypnum cupressiforme, 124
Hypochaeris radicata, 98